THE SIMPLER? POLYHEDRA

BEING THE THIRD PART
OF SEVERAL COMPRISING
THE COMPLETE? POLYHEDRA
BY PATRICK TAYLOR

THE SIMPLER? POLYHEDRA

First published 1999 by

Nattygrafix
277 Cavendish Street
Ipswich
Suffolk
IP3 8BQ

(01473) 289861

Made in England

ISBN 0 9516701 4 X

NATTYGRAFIX IPSWICH

CONTENTS

ILLUSTRATIONS

Figures * are supported by Tables nos. 1-29

for Russell Stannard, his enthusiasm and clarity

"...we have temporarily abandoned our hope of obtaining a proof that our enumeration is complete, but we shall be much surprised if any new uniform polyhedron is found in the future."

H S M Coxeter, M S Longuet-Higgins & J C P Miller *Uniform Polyhedra* 1954

INTRODUCTION

The question to be asked is what constitutes a uniform polyhedron?

This book is the third in a series: in the first, '*The Complete? Polygon*', we saw that compound polygons ({n/d} where n and d are not relatively prime) are no different in their derivation from other polygons, and also that double polygons sometimes result from truncation. In the second, '*Incomplete Tilings*', we saw that using these compound and double polygons did not preclude sensible results from being obtained: the normally accepted four basic families of tilings were extended by a further seven families of compound tilings without matters getting out of hand.

The same approach is now to be applied to the polyhedra, essentially assemblies of polygons, similar to tilings of the plane, but with fewer or lesser valued polygons at each vertex. This causes the two dimensional plane geometry of the tilings to be folded up into the three dimensional spherical geometry of the polyhedra.

The sheer quantity of material this generates will require two volumes to cover, so this third volume in the series will deal with the simpler polyhedra derived from the 'Platonic solids' and include an appendix with the dihedral symmetry families of the prisms. The fourth in the series will then explore the more complex super-families of the star and cross polyhedra.

It is the author's belief that including compound and double polygons, and hence compound and double polyhedra, will allow a more complete? picture to be painted of the various shapes possible; one without the curious gaps where certain truncations are deemed not to exist in the standard classification.

REGULAR POLYHEDRA

In the same way that a line can be divided into equal segments and bent around into a second dimension to form a regular polygon, a plane surface can be divided into the polygons of a regular tiling and then folded up into a regular convex figure that closes on itself by reducing the total angle of faces at each vertex to less than the plane's 360°.

Starting with the three regular tilings {3,6}, {4,4} and {6,3}, we can reduce the number of the polygons at each vertex to give {3,5}, {3,4}, {3,3} or {4,3}, or alternatively we can reduce the value of the polygons at each vertex to give {3,4}, {5,3}, {4,3} or {3,3}.

This gives us {5,3}, {4,3}, {3,3}, {3,4} and {3,5} as the possible ways of arranging polygons around vertices in a regular way.

These configurations give rise to the five Platonic solids known by the ancient Greeks: dodecahedron, cube (hexahedron), tetrahedron, octahedron and icosahedron respectively, as shown in Figure 1 opposite. Unlike the tilings which fill the plane to infinity, these are found to close properly and have finite numbers of vertices, edges and faces as shown in Table 1 below.

Their faces are all equal regular polygons, arranged identically around equal vertices, which can be described like the tilings in terms of 'dual edge lengths' making up a vertex figure. The vertex figures for these regular polyhedra are themselves regular polygons, such that the form {x,y} will have a vertex figure which is a regular y-gon, each face representing an x.

Symbol		V	E	F
3| 2 5 26 (5) Dodecahedron	{5,3}	20	30	12
3| 2 4 18 (3) Cube (Hexahedron)	{4,3}	8	12	6
3| 2 3 15 (1) Tetrahedron	{3,3}	4	6	4
4| 2 3 17 (2) Octahedron	{3,4}	6	12	8
5| 2 3 25 (4) Icosahedron	{3,5}	12	30	20

Table 1: Regular Polyhedra

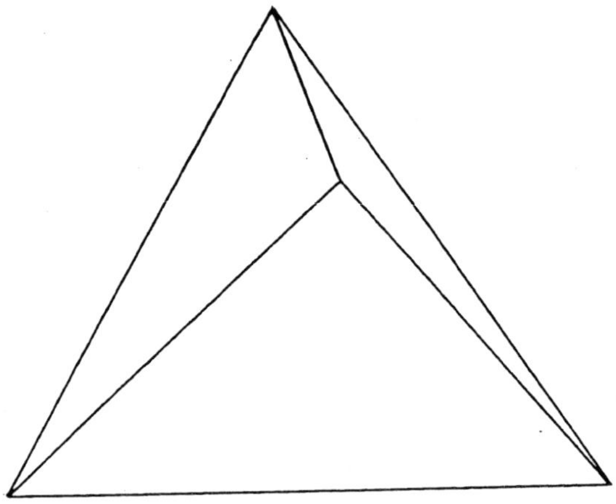

3| 2 3 {3,3}

15 (1)
Tetrahedron

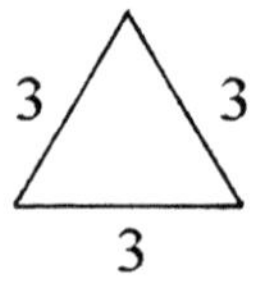

Fig.1: Regular Polyhedra

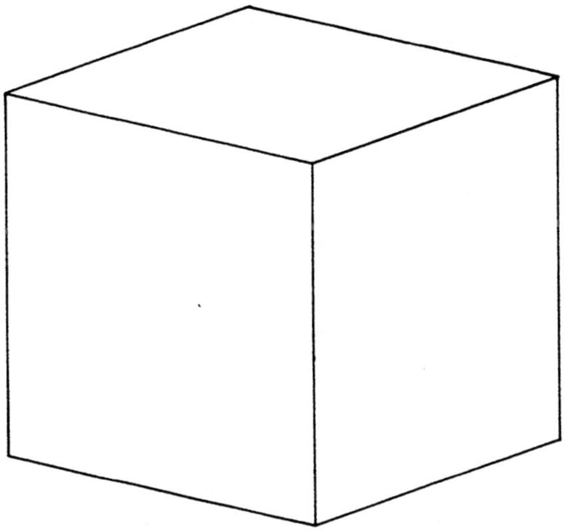

3| 2 4 {4,3}

18 (3)
Cube (Hexahedron)

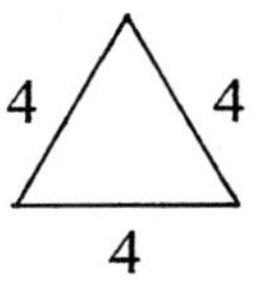

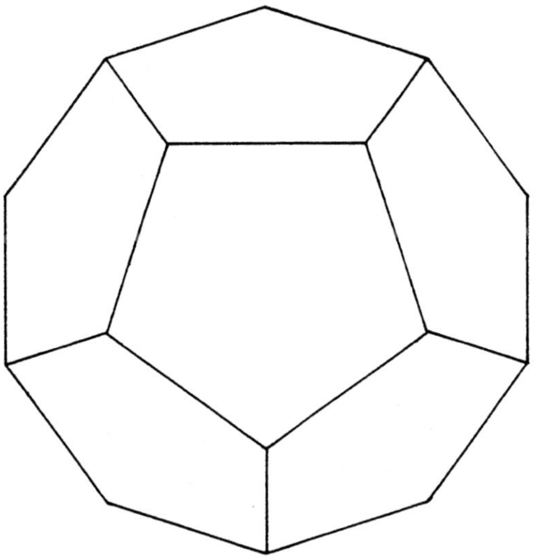

3| 2 5 {5,3}

26 (5)
Dodecahedron

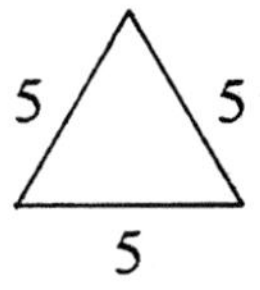

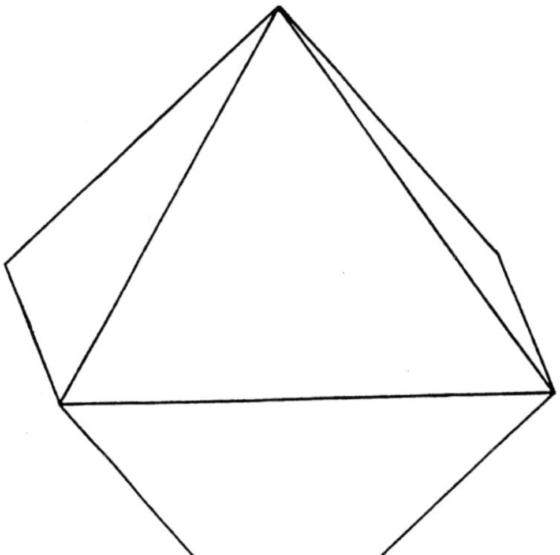

4| 2 3 {3,4}

17 (2)
Octahedron

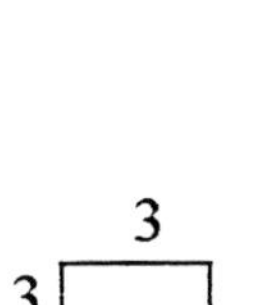

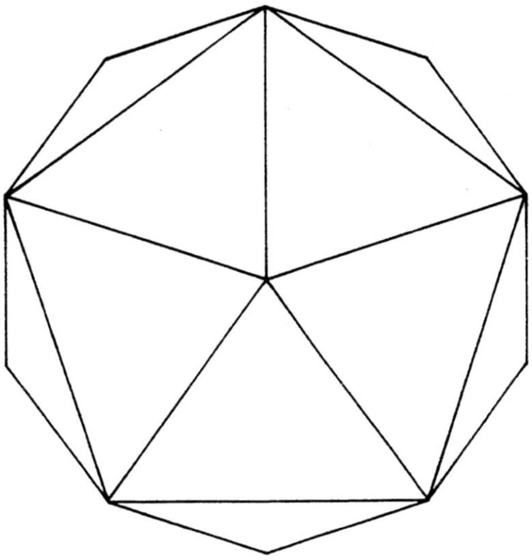

5| 2 3 {3,5}

25 (4)
Icosahedron

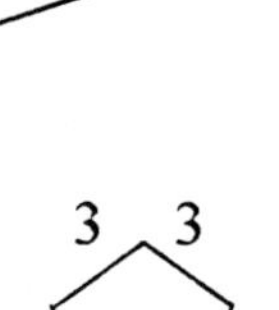

DIHEDRA AND DIGONS

Further reduction of the vertex or face values below those of the Platonic solids produces either two polygons back to back, i.e. {n,2}, a dihedron, or a collection of digons, {2,n}. These objects belong to an infinite collection of symmetry families that we shall deal with later in an appendix.

SOME NOTATION ETC.

Throughout this series we will adopt as standard the notation used by Coxeter et al. in their 1954 paper on '*Uniform Polyhedra*', e.g. y| 2 x, together with where applicable the slightly easier Schläfli notation, {x,y}.

In addition we shall include the numbering systems from that paper along with both Wenninger's excellently illustrative '*Polyhedron Models*' (-) and the author's own '*Additions to the Uniform Polyhedra*' [-], to provide useful cross references.

EULER'S FORMULA AND DENSITY

Looking again at Table 1 we can observe a property of the polyhedra which was first described in a formula by the great Swiss mathematician Leonhard Euler: V+F=E+2.

This will be found to hold for all the polyhedra, although later on for the more complex polyhedra to be encountered we will need to make adjustments on account of density. (Tilings have V+F=E on average)

We saw with the tilings that density was a measure of the number of layers of polygons involved in the tiling. With the polyhedra it will be seen to represent the number of layers of polygons we have to pass through to reach the centre of the polyhedron.

DUAL POLYHEDRA

From Table 1 we can see that the five Platonic solids seem to fall into three sets, two pairs where vertex and face symmetries are interchanged along with a single member with the same vertex and face symmetries.

The physical manifestation of these interchangeable symmetries can be seen by joining together the centres of the faces of any one of these polyhedra to form a dual polyhedron, generally the other of the pair: {x,y} becomes {y,x} as shown in Figure 2.

We thus find that our regular polyhedra naturally divide up into two dual pairs, {5,3} & {3,5} and {4,3} & {3,4}, together with one self-dual, {3,3}.

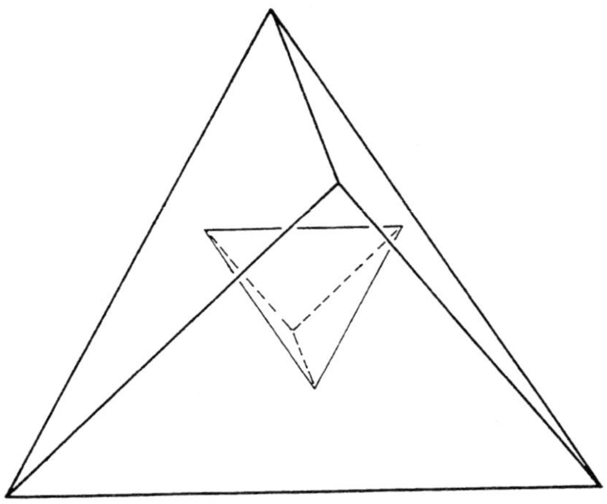

{3,3}, dual of {3,3}

Fig.2: Dual Polyhedra

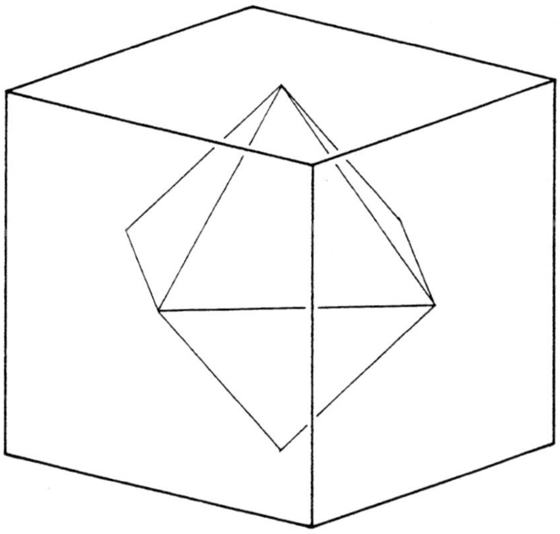

{3,4}, dual of {4,3}

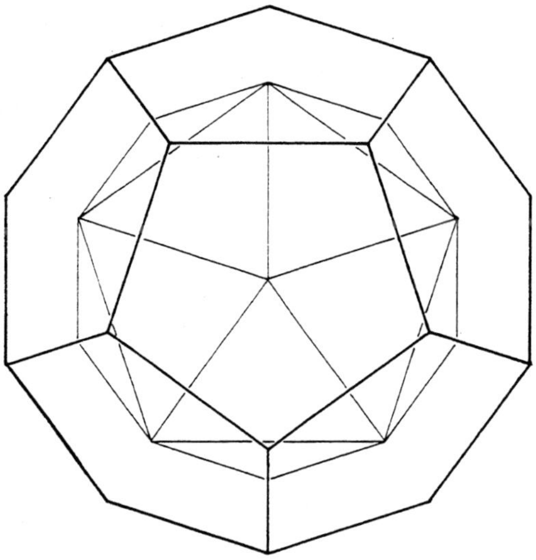

{3,5}, dual of {5,3}

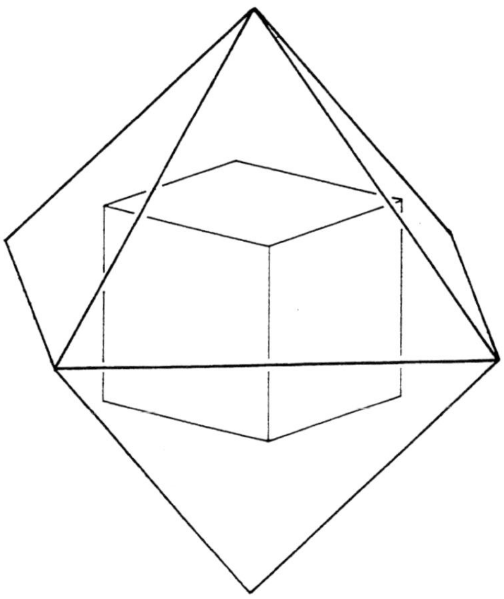

{4,3}, dual of {3,4}

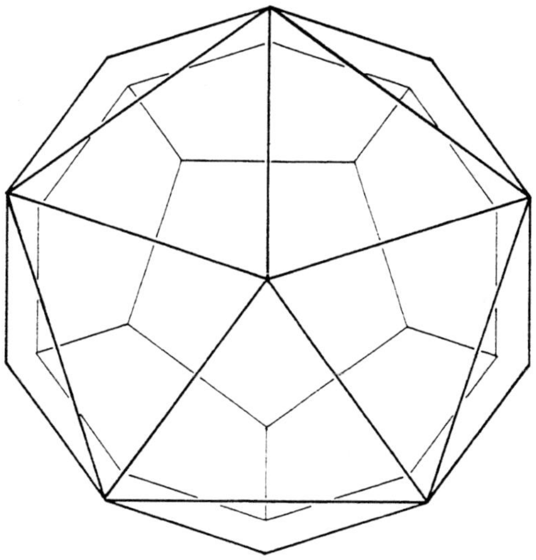

{5,3}, dual of {3,5}

TRUNCATED POLYHEDRA

The duality of these Platonic solids is face-vertex duality as opposed to the edge-vertex duality of a single polygon. From Figure 2 we can see that either of a dual pair can be obtained from the other by a process of truncating the vertices, removing everything until only the centres of the faces remain as the new vertices of the dual.

During this truncation process a number of further polyhedra are encountered, but semi-regular, consisting of regular polygons which are usually of two distinct types, but nevertheless assembled in a regular way.

The first of these truncations for each regular polyhedron is shown in Figure 3.

In general these have isosceles triangle vertex figures, consisting of one polygon for the truncating face, equivalent to the vertex figure polygon of the original parent solid, and two equal polygons formed from the truncated faces of that original. Thus {x,y} will have a truncation t{x,y} with one {y} and two {2x} polygons at each vertex, as shown below in Table 2.

These truncated polyhedra still have a single layer of polygons around the centre and thus a density of 1. In general for the simpler polyhedra the density will maintain this value of 1, whilst members of the sub-families considered towards the end of this volume will usually have a density of 2.

Symbol		V	E	F
3 2 |5	t{5,3}	60	90	12 ○
29 (10)				20 △
Truncated Dodecahedron				
3 2 |4	t{4,3}	24	36	6 ○
21 (8)				8 △
Truncated Cube				
3 2 |3	t{3,3}	12	18	4 ⬡
16 (6)				4 △
Truncated Tetrahedron				
4 2 |3	t{3,4}	24	36	8 ⬡
20 (7)				6 □
Truncated Octahedron				
5 2 |3	t{3,5}	60	90	20 ⬡
27 (9)				12 ⬠
Truncated Icosahedron				

Table 2: Truncated Polyhedra

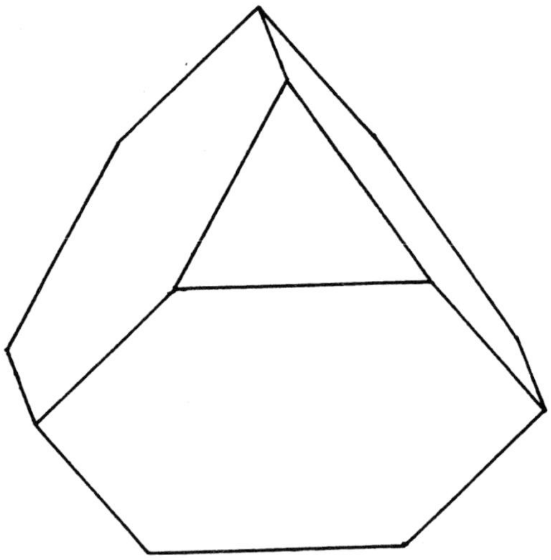

3 2 |3 t{3,3}

16 (6)

Truncated Tetrahedron

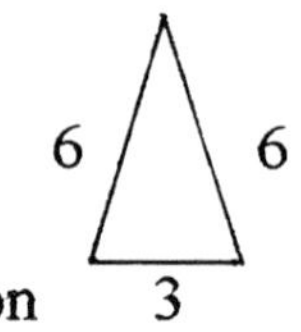

Fig.3: Truncated Polyhedra

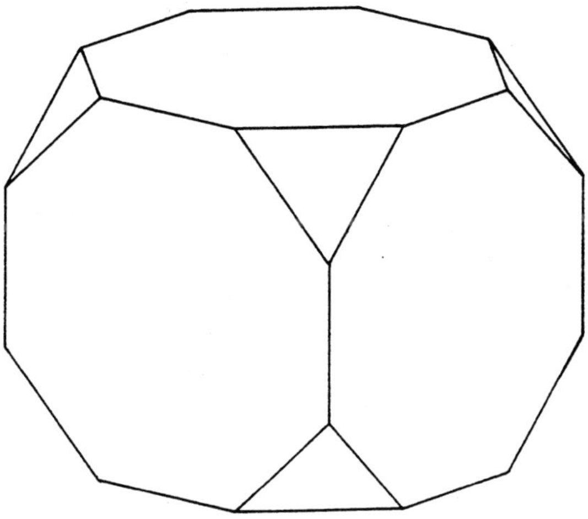

3 2 |4 t{4,3}

21 (8)
Truncated Cube

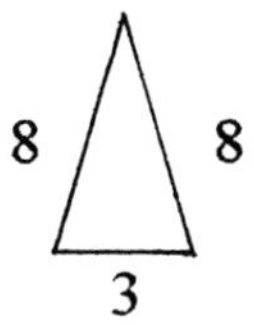

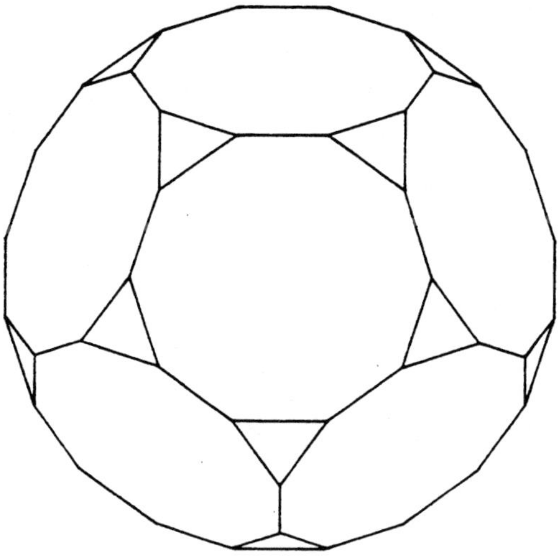

3 2 |5 t{5,3}

29 (10)
Truncated Dodecahedron

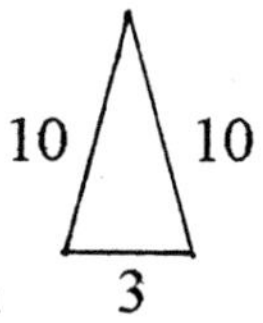

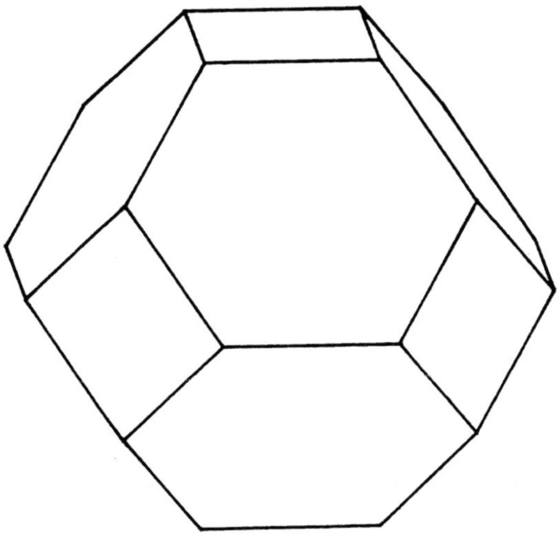

4 2 |3 t{3,4}

20 (7)
Truncated Octahedron

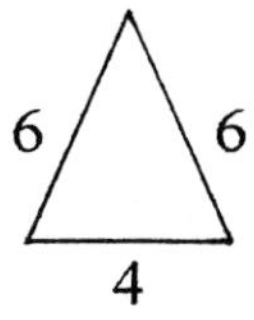

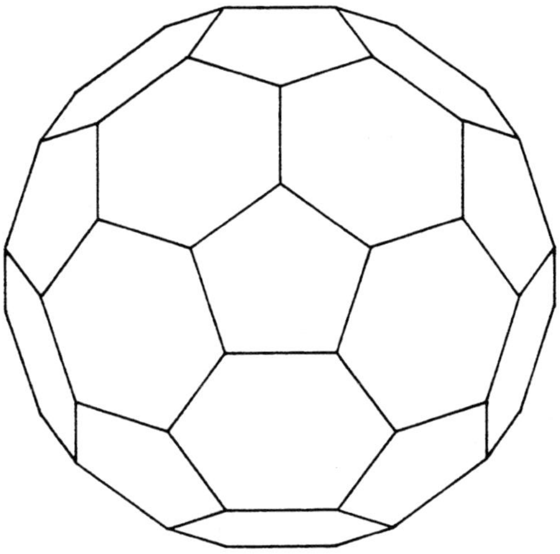

5 2 |3 t{3,5}

27 (9)
Truncated Icosahedron

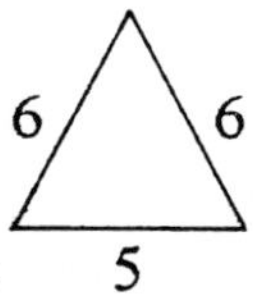

QUASI-REGULAR POLYHEDRA

Continuing the truncation process beyond the first truncation of a regular polyhedron leads on to a second truncation. This is reached at the point when the original polygon faces of the parent solid have become their own duals and the truncating faces meet at their vertices. All faces are again regular, and generally of two types, one from each parent solid.

These faces are arranged alternating around a rectangular vertex figure, and as we found with the tilings, the second truncations of either of a dual pair are in fact the same semi-regular polyhedron. Both {x,y} and {y,x} lead to a polyhedron with {x} and {y} faces alternating around the vertices.

These quasi-regular polyhedra are illustrated in Figure 4 and described in Table 3 where it can be seen that Euler's formula, V+F=E+2 still holds good.

Interestingly this truncation process produces only two new polyhedra, one each for the {3,4} and {3,5} symmetry families. The equivalent polyhedron for the self dual {3,3} family has two triangles and two triangles at each vertex, i.e. four triangles, and is consequently the already seen Octahedron.

Symbol		V	E	F
2| 3 5 28 (12) Icosidodecahedron	{5} {3}	30	60	12 ⬠ 20 △
2| 3 4 19 (11) Cuboctahedron	{4} {3}	12	24	6 □ 8 △
2| 3 3 17 (2) Octahedron	{3} {3}	6	12	4 △ 4 △
2| 4 3 19 (11) Cuboctahedron	{3} {4}	12	24	8 △ 6 □
2| 5 3 28 (12) Icosidodecahedron	{3} {5}	30	60	20 △ 12 ⬠

Table 3: Quasi-Regular Polyhedra

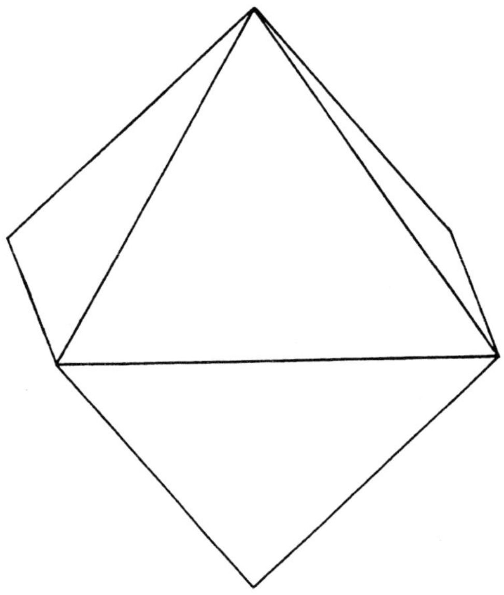

2| 3 3 {3 3}

17 (2)

Octahedron

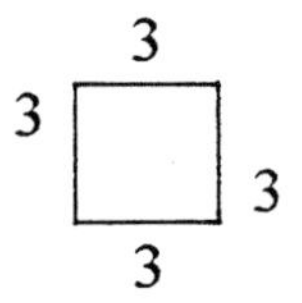

Fig.4: Quasi-Regular Polyhedra

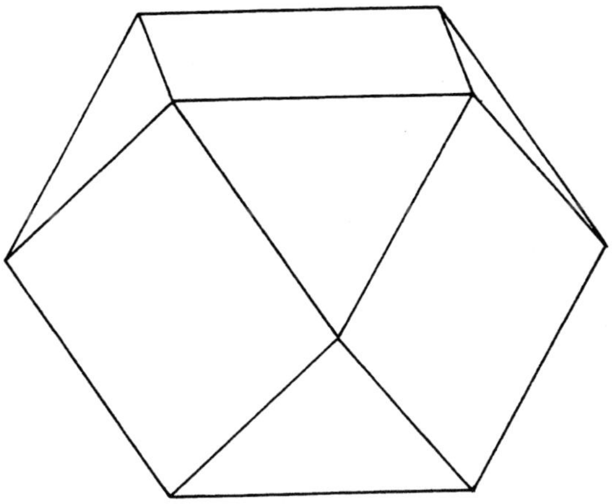

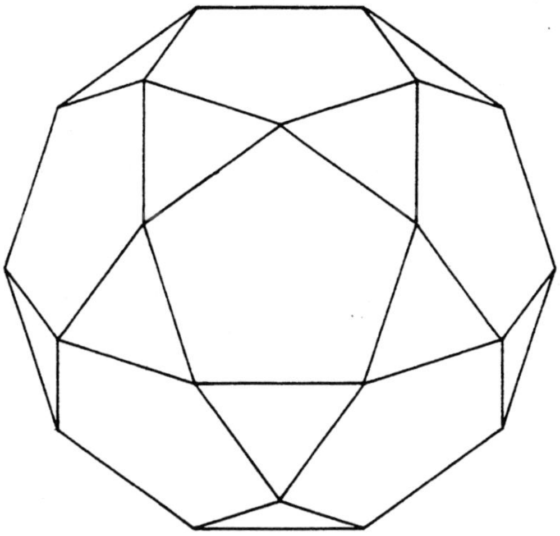

2| 3 4 $\begin{Bmatrix}4\\3\end{Bmatrix}$

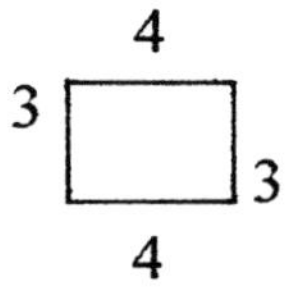

19 (11)
Cuboctahedron

2| 3 5 $\begin{Bmatrix}5\\3\end{Bmatrix}$

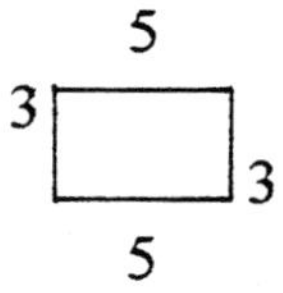

28 (12)
Icosidodecahedron

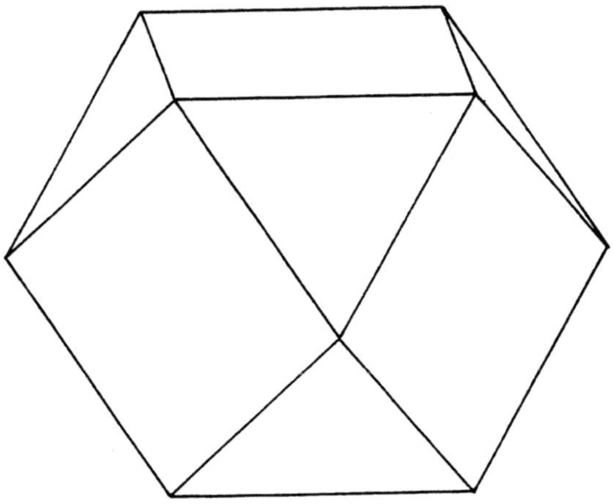

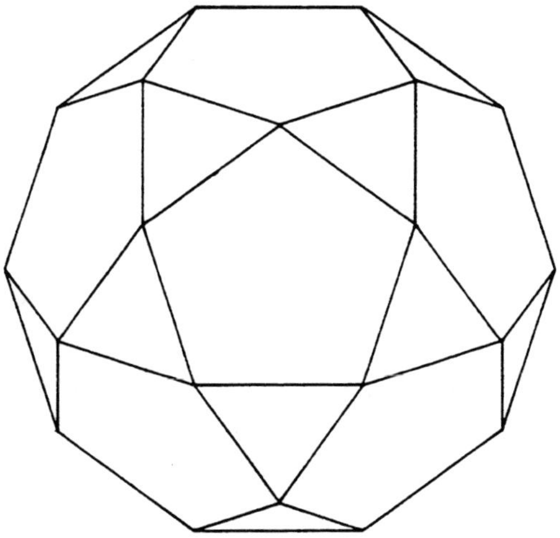

2| 4 3 $\begin{Bmatrix}3\\4\end{Bmatrix}$

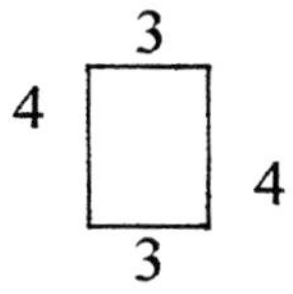

19 (11)
Cuboctahedron

2| 5 3 $\begin{Bmatrix}3\\5\end{Bmatrix}$

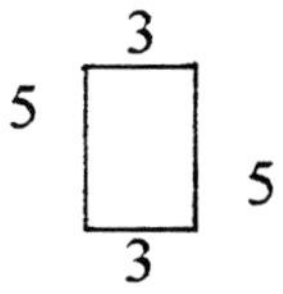

28 (12)
Icosidodecahedron

FIRST TRUNCATION SERIES

The quasi-regular polyhedron for each dual pair can thus be seen as the centre of a truncation series that leads from one of the pair to the other.

Continued truncation beyond this leads to a third semi-regular truncation where the truncating faces have expanded further and truncated each other, this third truncation of one of a dual pair being the first truncation, already seen, of the other of the pair.

Eventually by truncating even further, all that will remain is the centre of each original face, leaving us with the dual itself.

These truncation series for each symmetry family are shown in Figure 5. They are each composed of five polyhedra, but contain within them two shorter truncation series of three polygons each, overlapped at the central quasi-regular polyhedron:

(x	2x	x)	x	.
.	y	(y	2y	y)

{x,y} ↔ {y,x}

We will find in our later work, in a future volume on the closepacks and hypersolids, that they in turn form truncation series seven figures long. These will be seen to be the result of solid-vertex duality and found to consist of two polyhedron series overlapped at three central figures which share a polygon truncation series:

(x	2x	x	x	.)	.	.
(.	y	y	2y	y)	y	.
.	y	(y	2y	y	y	.)
.	.	(.	z	z	2z	z)

{x,y,z} ↔ {z,y,x}

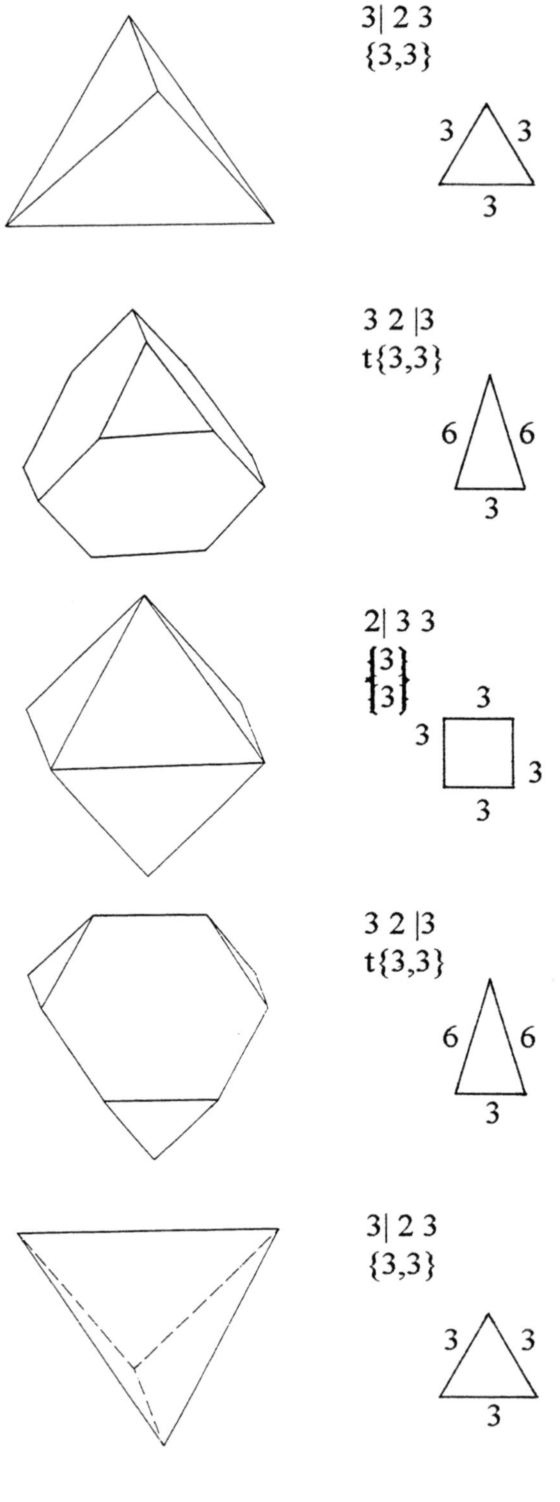

Fig.5: First Series, {3,3} Family

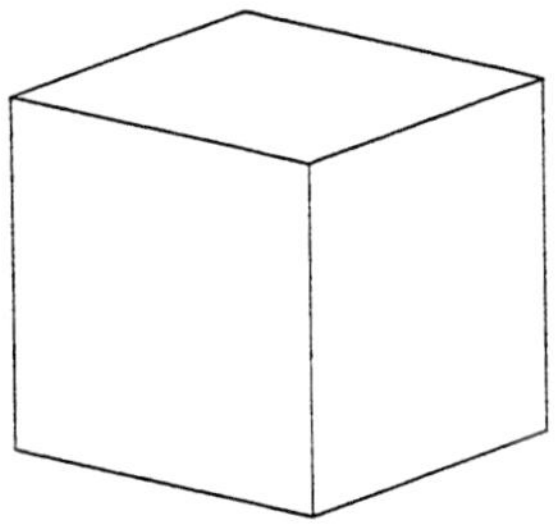

3| 2 4
{4,3}

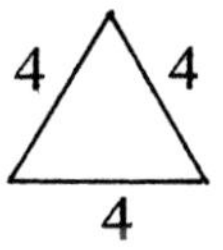

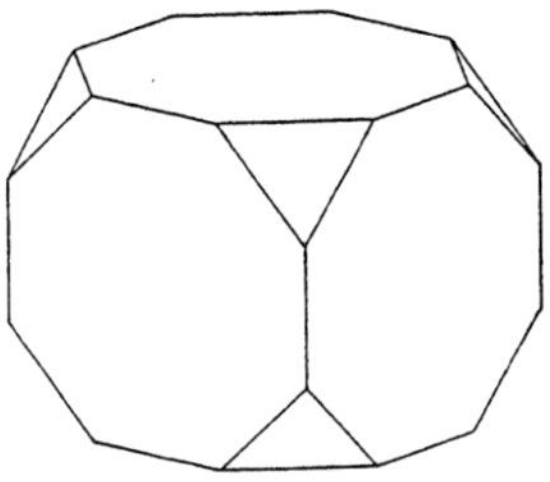

3 2 |4
t{4,3}

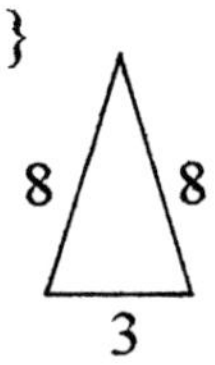

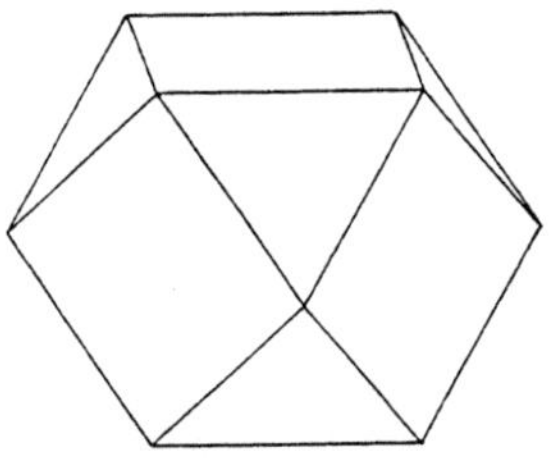

2| 3 4
$\begin{Bmatrix}3\\4\end{Bmatrix}$

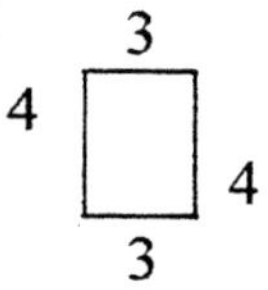

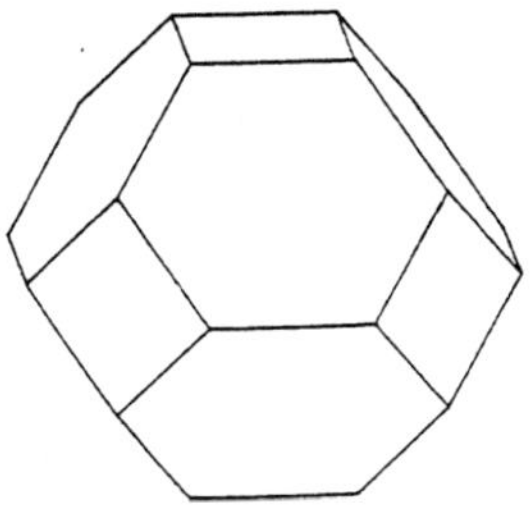

4 2 |3
t{3,4}

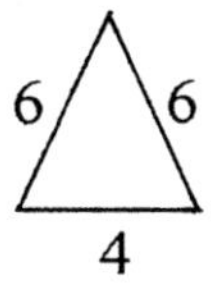

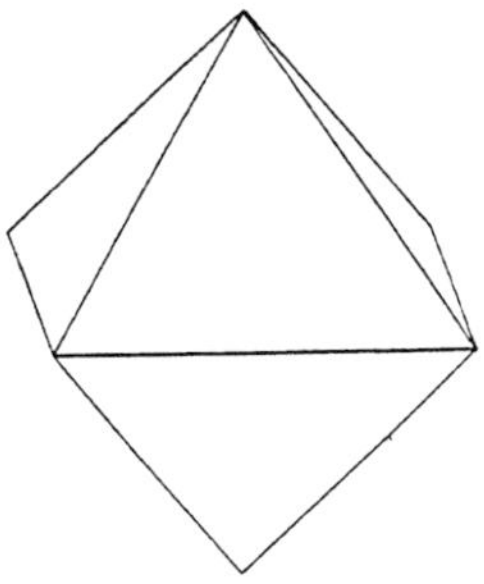

4| 2 3
{3,4}

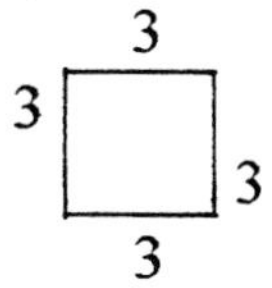

First Series, {3,4} Family

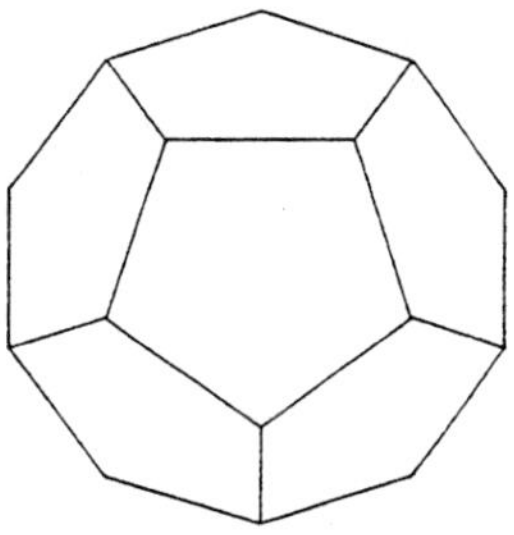

3| 2 5
{5,3}

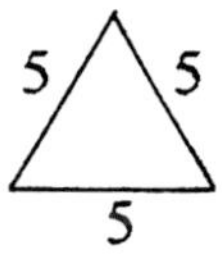

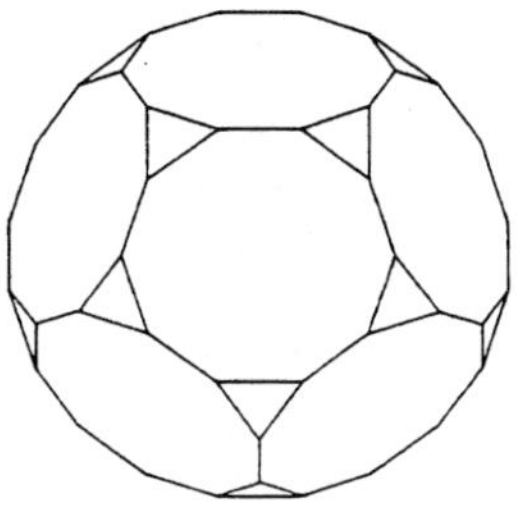

3 2 |5
t{5,3}

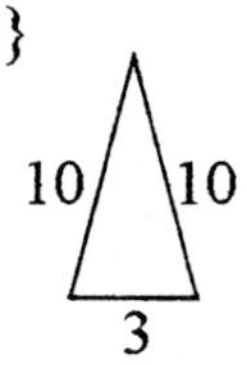

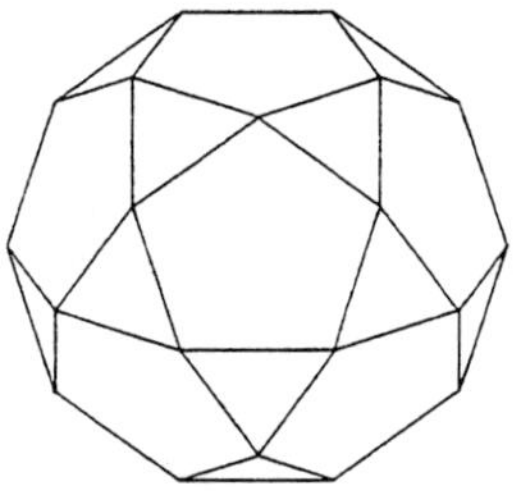

2| 3 5
$\begin{Bmatrix}3\\5\end{Bmatrix}$

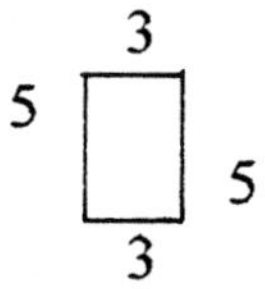

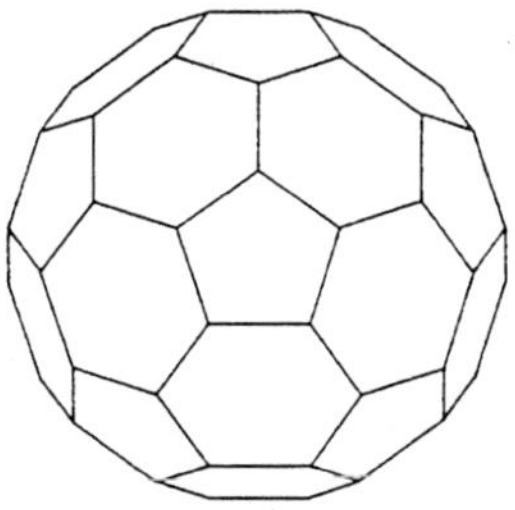

5 2 |3
t{3,5}

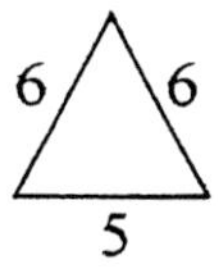

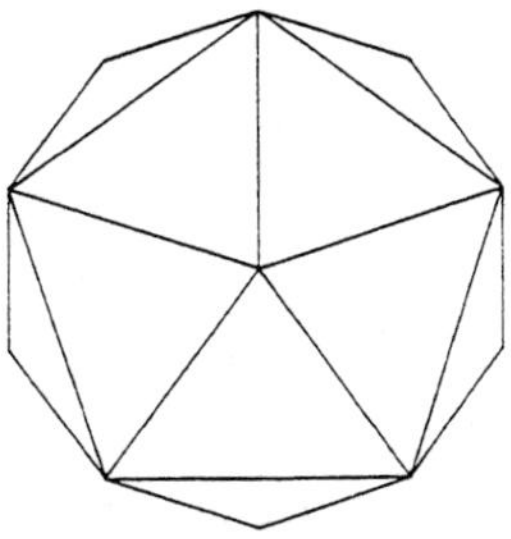

5| 2 3
{3,5}

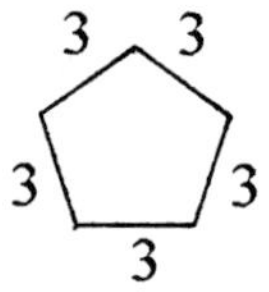

First Series, {3,5} Family

RHOMBIC POLYHEDRA

We have already seen that the self-dual {3,3} family has an Octahedron as its quasi-regular polyhedron, and we know this can be truncated as the first series of the {3,4} family to produce a Cube as its dual.

The two other quasi-regular polyhedra we have seen also have duals. Joining the centres of the faces of these we can produce a 'false' dual composed of twice as many isosceles triangles as the quasi-regular polyhedron has vertices. When this process is repeated it does not yield back the original quasi-regular polyhedron , but rather something irregular with twice as many vertices as we started with.

We can however produce a 'true' dual by simply truncating the vertices of the quasi-regular polyhedra. This produces at first a new rectangular truncating face, which gradually grows and merges with its fellows to become a rhombic face. These rhombic polyhedra have two types of vertices and are shown in Figure 6 along with their duals, the quasi-regular polyhedra, formed by either joining the centres of the rhombic faces or by truncating them in the usual manner.

The reason we get a 'false' dual in one case and rectangular faces in the other is that the two different types of face in the quasi-regular polyhedra are not equidistant from the centre, only their shared edges are. The rhombic faces of the rhombic polyhedra have diagonal lengths proportional to the dual edge lengths of the two polygons comprising the quasi-regular polyhedron. These diagonals are shown in Figure 7, where they can be seen to delineate the two parent solids compounded. That from the {3,3} family is Kepler's Stella Octangula, a uniform polyhedron in its own right (qv).

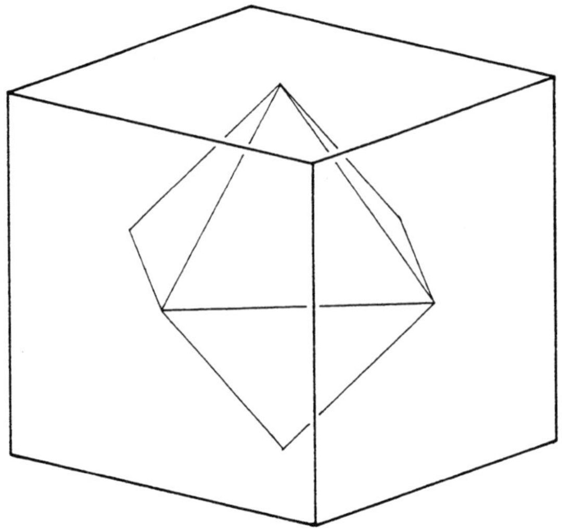

$\left\{\begin{matrix}3\\3\end{matrix}\right\}$, dual of Cube

Fig.6: Duals of Rhombic Polyhedra

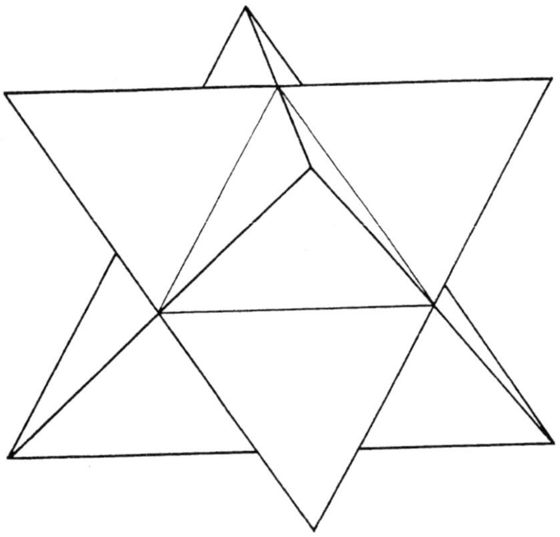

{3,3} and {3,3}

(19) [1.4a]
Stella Octangula

Fig.7: Diagonals of Rhombic Polyhedra

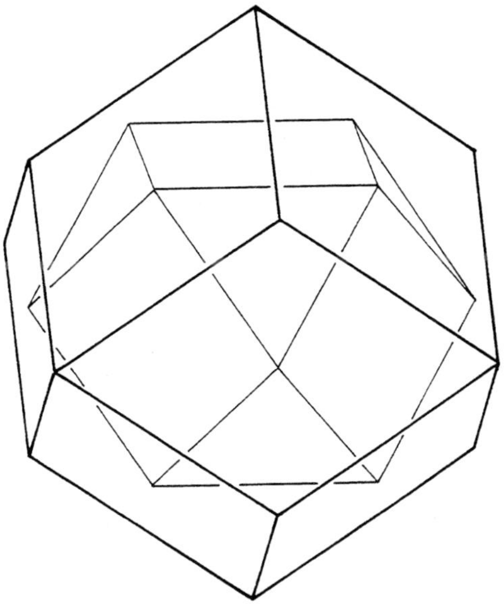

$\begin{Bmatrix}3\\4\end{Bmatrix}$, dual of Rhombic Dodecahedron

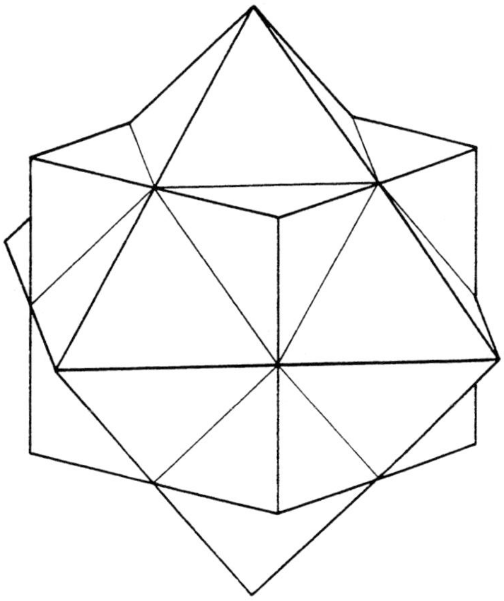

{3,4} and {4,3}

(43)
Octahedron and Cube

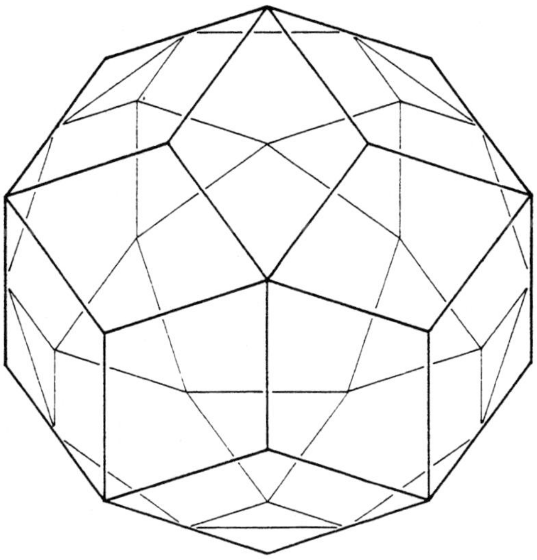

$\begin{Bmatrix}3\\5\end{Bmatrix}$, dual of Rhombic Triacontahedron

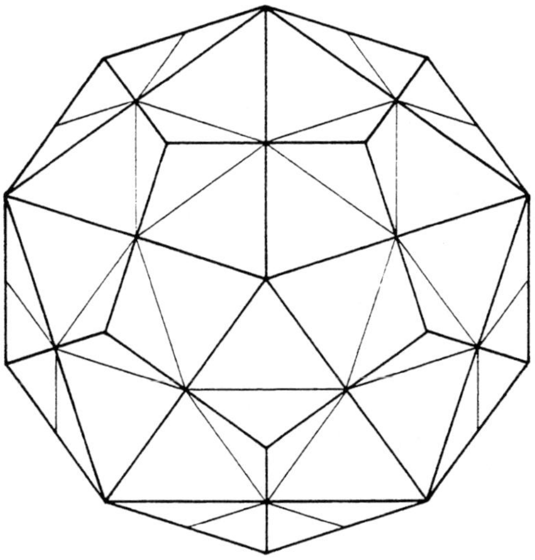

{3,5} and {5,3}

(47)
Icosahedron and Dodecahedron

EVEN FACED POLYHEDRA

We have just seen that truncating the quasi-regular polyhedra in general produces a rectangular face. We can however normalise these rectangles, as we did with the tilings, to produce semi-regular polyhedra that seem to be truncations of the quasi-regulars.

The first 'truncation' of a quasi-regular will therefore consist of the truncated faces of the two parent polyhedra, {2x} and {2y}, along with a number of squares {4} formed from the rectangles. These polyhedra are thus called even faced and have triangular vertex figures as shown in Figure 8.

Although not true truncations of the quasi-regulars they are true truncations of the rhombic polyhedra, when that truncation is carried out at different rates at the two vertex types.

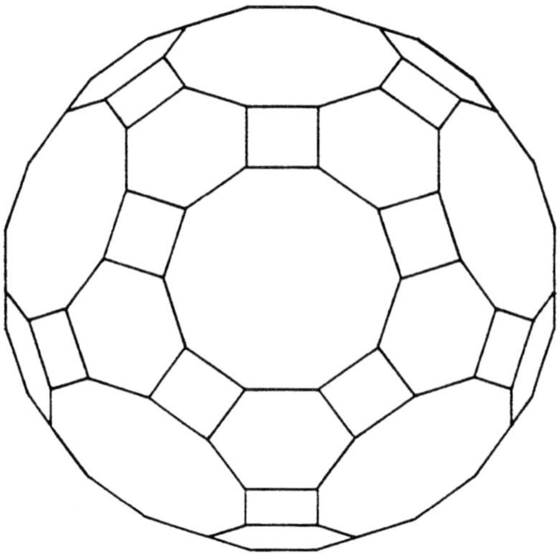

3 2 5| $t\begin{Bmatrix}3\\5\end{Bmatrix}$

31 (16)

Truncated Icosidodecahedron

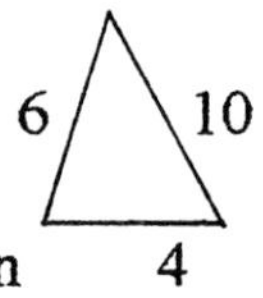

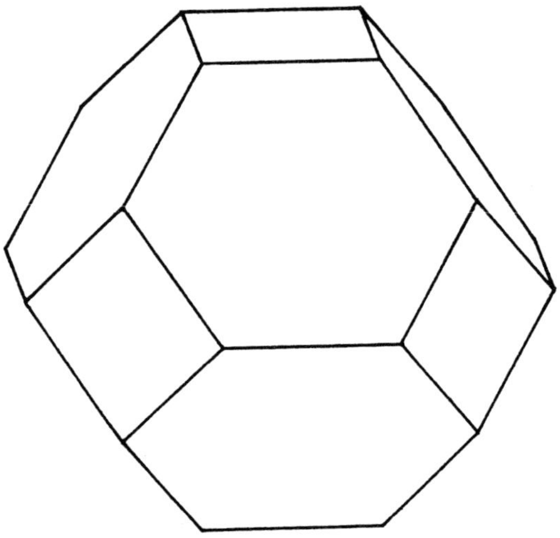

3 2 3| $t\begin{Bmatrix}3\\3\end{Bmatrix}$

20 (7)

Truncated Octahedron

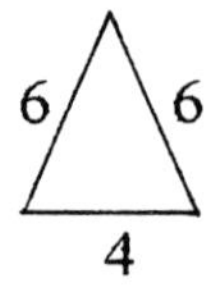

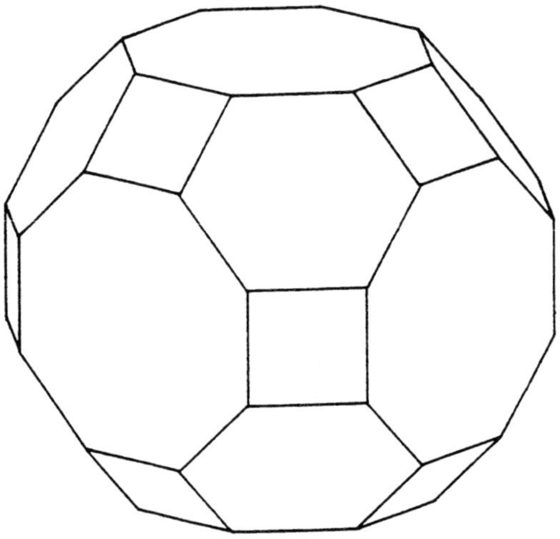

3 2 4| $t\begin{Bmatrix}3\\4\end{Bmatrix}$

23 (15)

Truncated Cuboctahedron

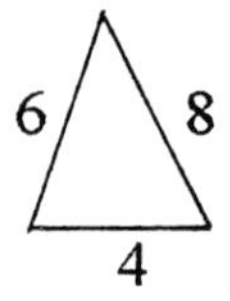

Fig.8: Even Faced Polyhedra

RHOMBIHEDRA

The second 'truncation' of a rhombic polyhedron can similarly be equated to a normalised second truncation of the dual quasi-regular polyhedron.

These semi-regular rhombihedra generally have isosceles trapezoid vertex figures, the two parent polyhedra providing the opposed parallel sides, {x} and {y}, and two opposed squares {4} joining them together.

The rhombihedron for each symmetry family is shown in Figure 9, where as we saw with the even faced polyhedra, the representative of the {3,3} family has already been seen in the first truncation series of the {3,4} family.

The even faced polyhedra, rhombihedra and snub polyhedra shown overleaf are described numerically in Tables 4, 5 and 6 that follow.

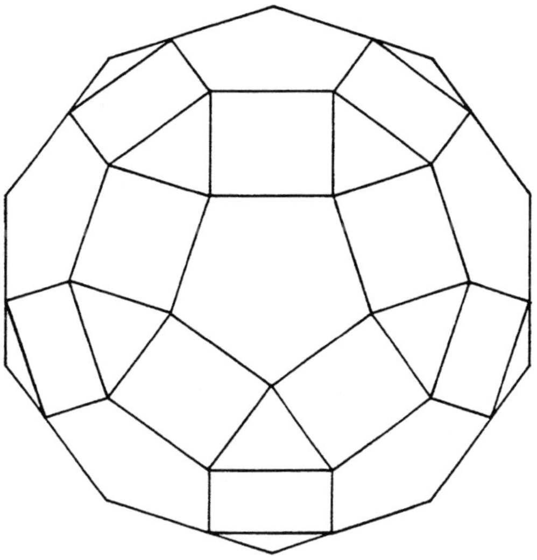

3 5 |2 $r\begin{Bmatrix}3\\5\end{Bmatrix}$

30 (14)

Rhombicosidodecahedron

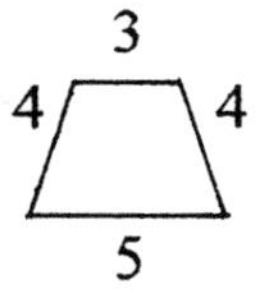

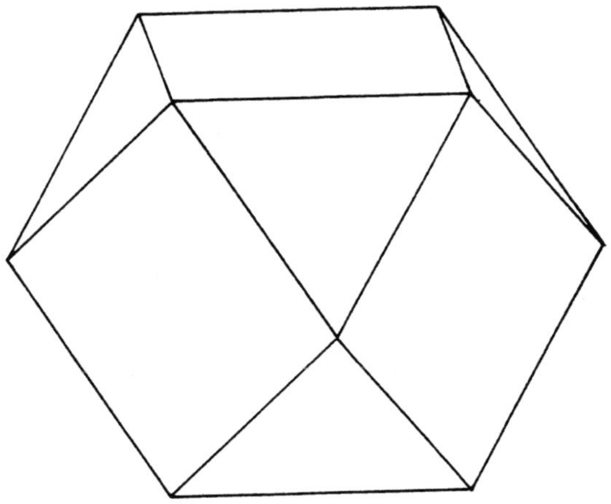

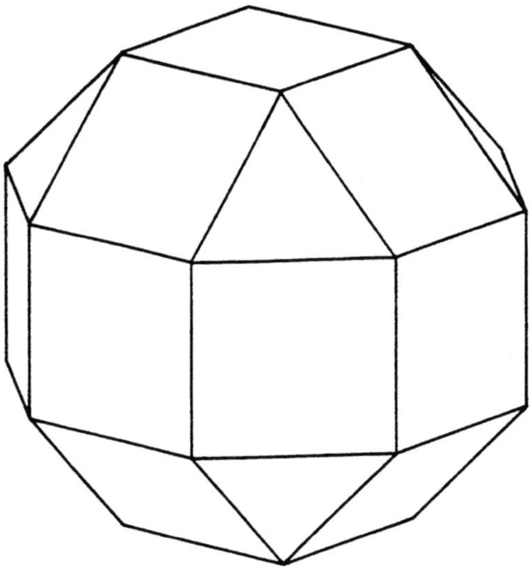

3 3 |2 $r\begin{Bmatrix}3\\3\end{Bmatrix}$

19 (11)

Cuboctahedron

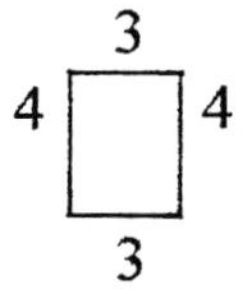

3 4 |2 $r\begin{Bmatrix}3\\4\end{Bmatrix}$

22 (13)

Rhombicuboctahedron

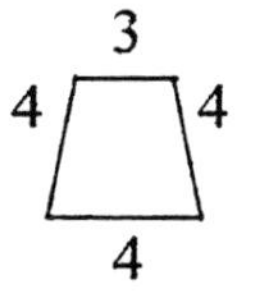

Fig.9: Rhombihedra

SNUB POLYHEDRA

One further uniform polyhedron remains for each symmetry family, which can be produced by taking the rhombihedron, omitting the truncating square faces and twisting the remainder to leave room for two equilateral triangle faces in their place.

These are the snub polyhedra shown in Figure 10, that for the {3,3} family again having been seen before, but this time from the {3,5} family in the form of the Icosahedron. The other two occur in two enantiomorphic mirror image forms differentiated by the direction of the twist.

This brings our convex uniform polyhedra up to a full complement of 18, the 5 regular Platonic solids and 13 semi-regulars known as the Archimedean solids.

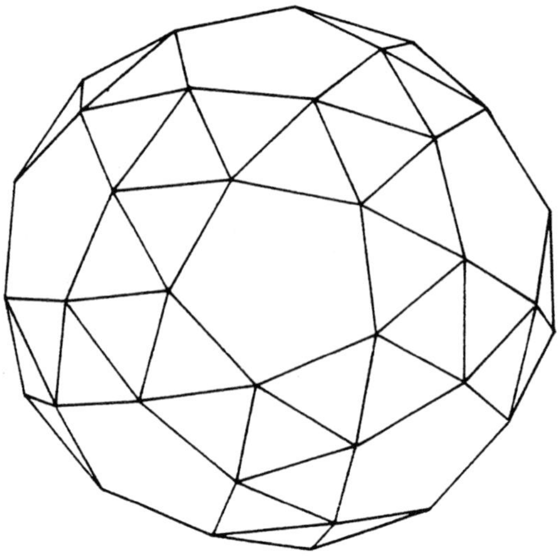

|3 2 5 $s\begin{Bmatrix}3\\5\end{Bmatrix}$

32 (18)
Snub Dodecahedron

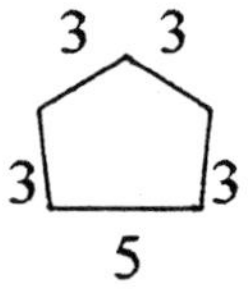

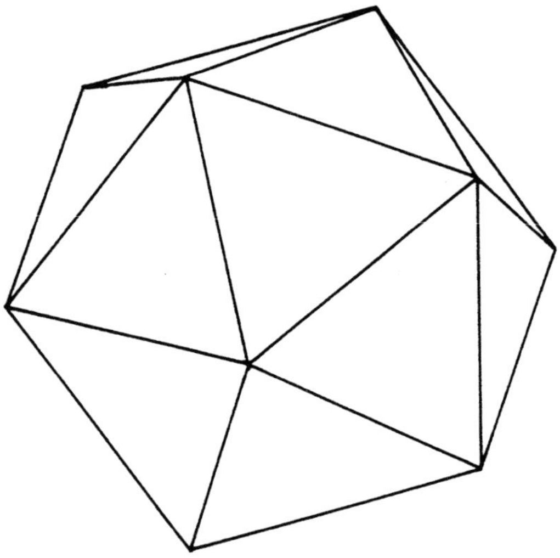

|3 2 3 $s\begin{Bmatrix}3\\3\end{Bmatrix}$

25 (4)
Snub Tetrahedron

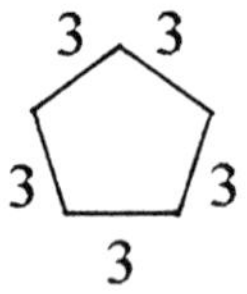

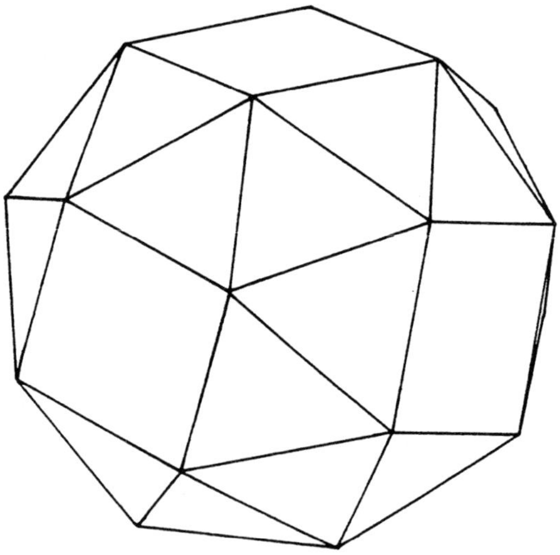

|3 2 4 $s\begin{Bmatrix}3\\4\end{Bmatrix}$

24 (17)
Snub Cube

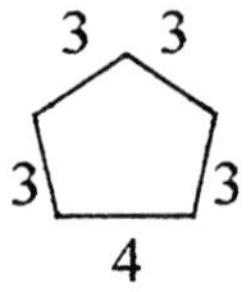

Fig.10: Snub Polyhedra

Symbol		V	E	F
3 2 5|	$t\begin{Bmatrix}3\\5\end{Bmatrix}$	120	180	12 ○
31 (16)				20 ⬡
Truncated Icosidodecahedron				30 □
3 2 4|	$t\begin{Bmatrix}3\\4\end{Bmatrix}$	48	72	6 ○
23 (15)				8 ⬡
Truncated Cuboctahedron				12 □
3 2 3|	$t\begin{Bmatrix}3\\3\end{Bmatrix}$	24	36	4 ⬡
20 (7)				4 ⬡
Truncated Octahedron				6 □

Table 4: Even Faced Polyhedra

3 5 |2	$r\begin{Bmatrix}3\\5\end{Bmatrix}$	60	120	12 ⬠
30 (14)				20 △
Rhombicosidodecahedron				30 □
3 4 |2	$r\begin{Bmatrix}3\\4\end{Bmatrix}$	24	48	6 □
22 (13)				8 △
Rhombicuboctahedron				12 □
3 3 |2	$r\begin{Bmatrix}3\\3\end{Bmatrix}$	12	24	4 △
19 (11)				4 △
Cuboctahedron				6 □

Table 5: Rhombihedra

|3 2 5	$s\begin{Bmatrix}3\\5\end{Bmatrix}$	60	150	12 ⬠
32 (18)				20 △
Snub Dodecahedron				60 △
|3 2 4	$s\begin{Bmatrix}3\\4\end{Bmatrix}$	24	60	6 □
24 (17)				8 △
Snub Cube				24 △
|3 2 3	$s\begin{Bmatrix}3\\3\end{Bmatrix}$	12	30	4 △
25 (4)				4 △
Snub Tetrahedron				12 △

Table 6: Snub Polyhedra

MÖBIUS TRIANGLES

The rhombs of the rhombic polyhedra can each be broken down into four congruent triangles, on the edges and bisectors of which the vertices of all but the snub polyhedra can be readily found.

A typical example of such a Möbius triangle is shown in Figure 11, where each vertex is labelled with the symmetry that occurs there, i.e. x, y or 2.

By convention in Coxeter's notation, the various points around each triangle that form the vertices of the other members of the same symmetry family are then labelled.

Regular and quasi-regular polyhedra occur at the vertices of the Möbius triangle, truncated polyhedra and rhombihedra occur on intersections of edges and bisectors and the even faced polyhedron occurs in the centre where all three bisectors meet.

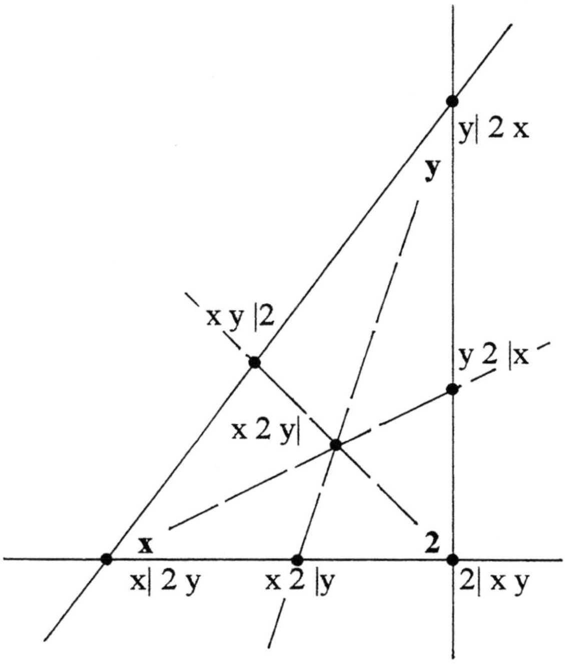

Fig.11: Typical Möbius Triangle

A SECOND SERIES

As with the tilings it is not possible to produce a notional third 'truncation' of the quasi-regular polyhedron except for the self-dual {3,3} family. To do so would require the normalising of a truncated rhomb into a regular octagon. Whilst two such octagons and a triangle can form a vertex, as in the {3,3} family, two of them and a square produces a tiling, whilst two plus a pentagon has too large a total angle at the vertex and is not at all feasible.

We can however still produce a second series of five polyhedra for each dual pair, leading from the quasi-regular to the rhombic, by introducing the snub polyhedron to conveniently fill the impossible gap.

This second series thus comprises a quasi-regular polyhedron (with a rectangular vertex figure), its 'truncation' the even faced polyhedron (with a triangular vertex figure), its second 'truncation' the rhombihedron (with an isosceles trapezium vertex figure), the snub polyhedron obtained by twisting the vertices of this last, and finally the rhombic polyhedron itself, dual of the quasi-regular polyhedron and whose two sets of diagonals delineate the original dual pair.

Figure 12 shows this second series for each of our three symmetry families {3,3}, {3,4} and {3,5}.

We have now considered and described all 18 of the regular and semi-regular polyhedra which consist of a single spherical layer of edge to edge polygons. These polygons have individually had a density of 1, which value can be assigned to each polyhedron in turn as a measure of the number of layers that have to be crossed to reach the centre.

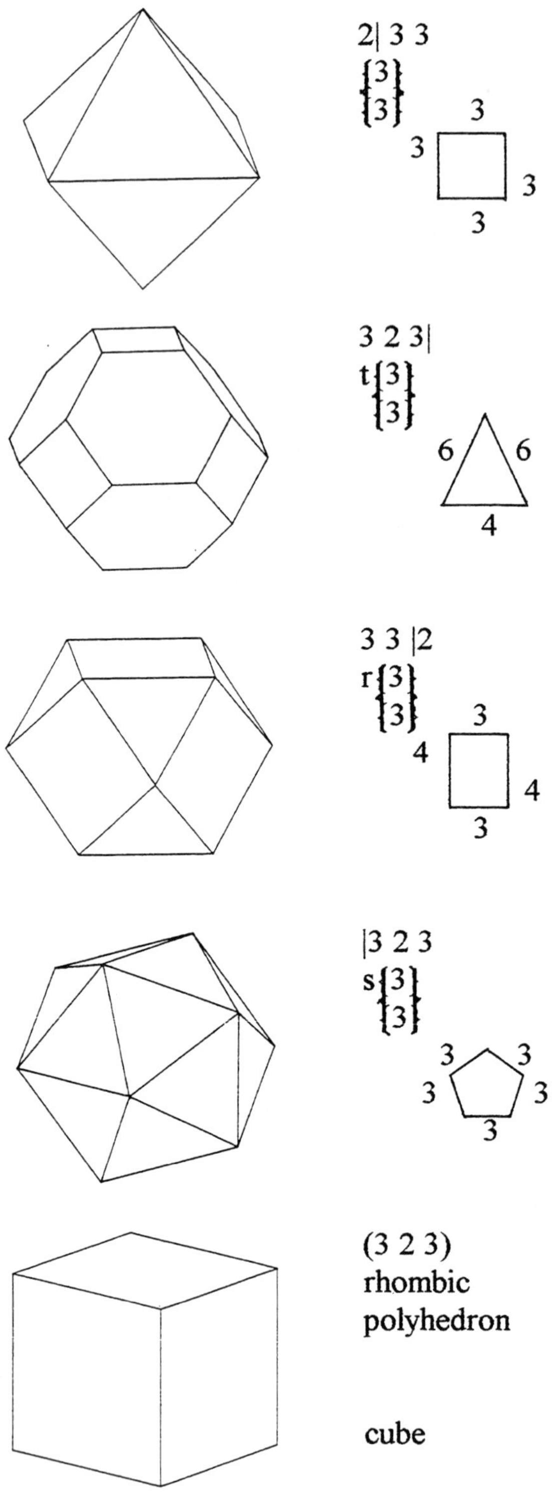

Fig.12: Second Series, {3,3} Family

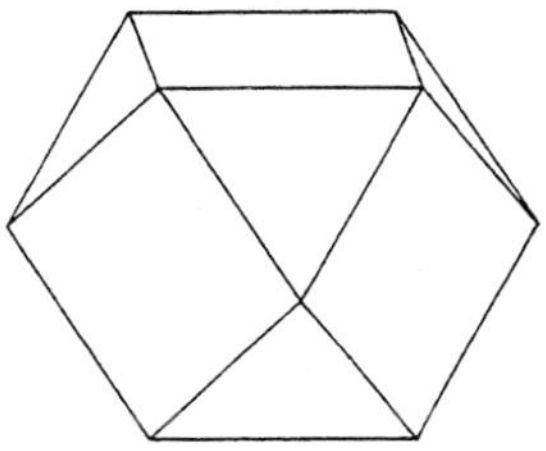

2| 3 4

$\begin{Bmatrix}3\\4\end{Bmatrix}$

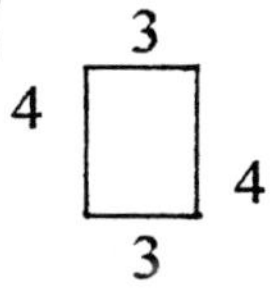

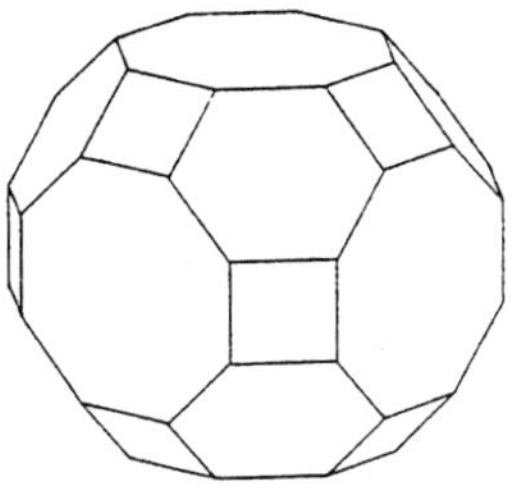

3 2 4|

$t\begin{Bmatrix}3\\4\end{Bmatrix}$

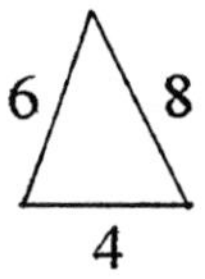

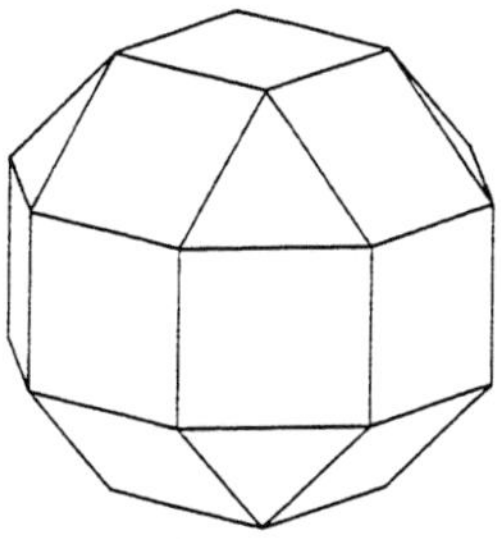

3 4 |2

$r\begin{Bmatrix}3\\4\end{Bmatrix}$

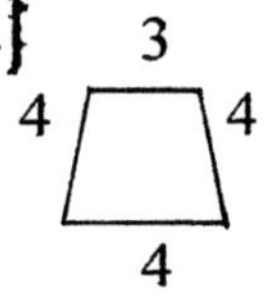

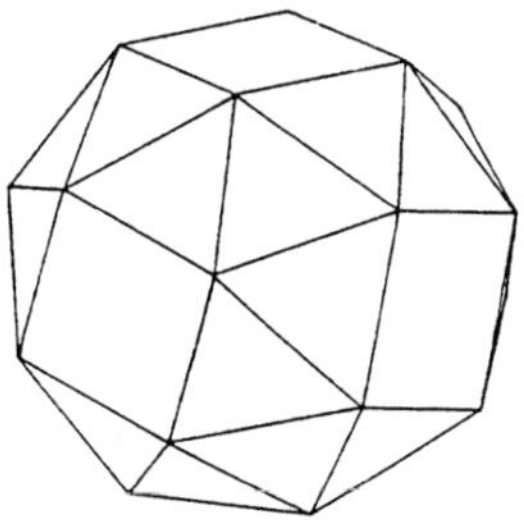

|3 2 4

$s\begin{Bmatrix}3\\4\end{Bmatrix}$

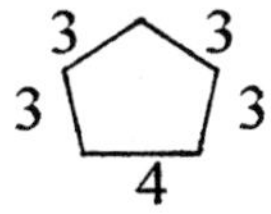

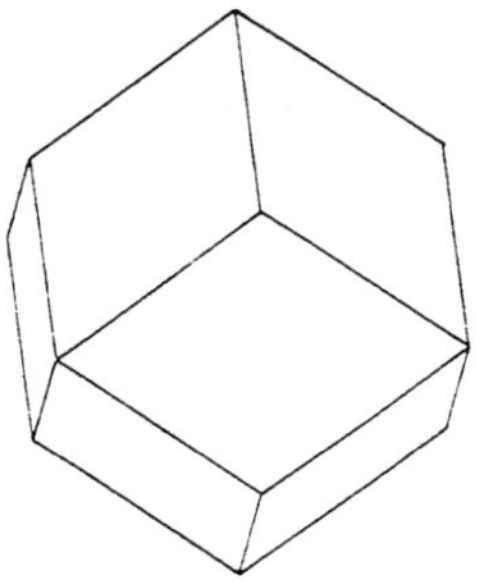

(3 2 4)
rhombic
polyhedron

rhombic
dodecahedron

Second Series, {3,4} Family

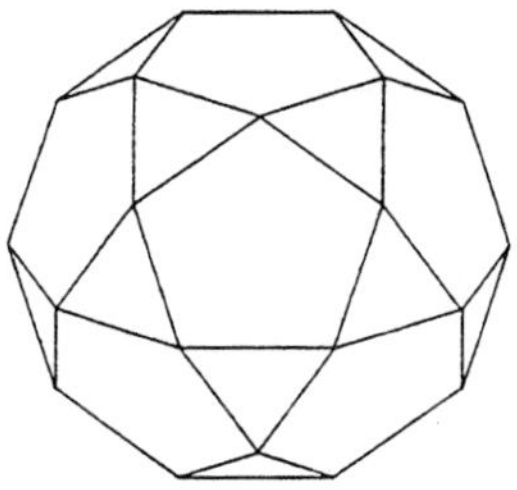

2| 3 5

$\begin{Bmatrix}3\\5\end{Bmatrix}$

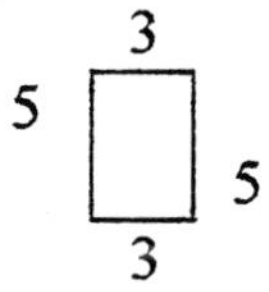

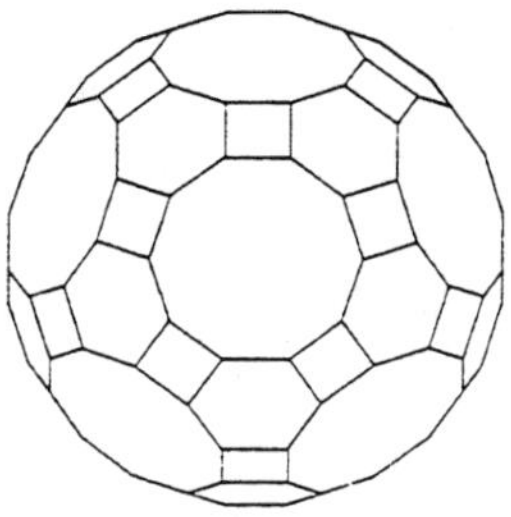

3 2 5|

$t\begin{Bmatrix}3\\5\end{Bmatrix}$

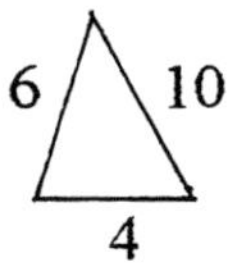

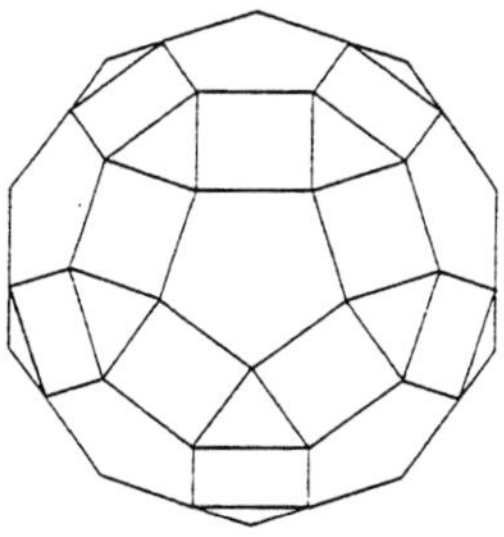

3 5 |2

$r\begin{Bmatrix}3\\5\end{Bmatrix}$

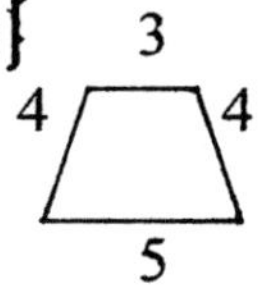

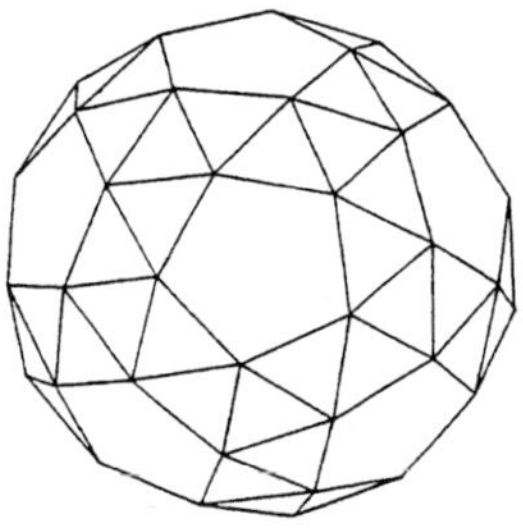

|3 2 5

$s\begin{Bmatrix}3\\5\end{Bmatrix}$

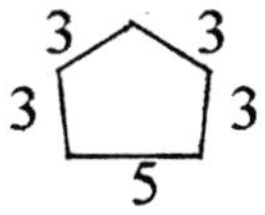

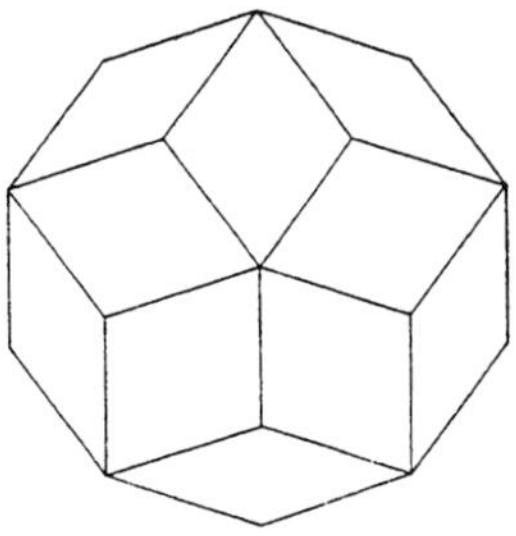

(3 2 5)
rhombic
polyhedron

rhombic
triacontahedron

Second Series, {3,5} Family

SPHERICAL KALEIDOSCOPES

The Möbius triangles described on page 21 form the constituent parts of a spherical kaleidoscope for each symmetry family as shown in Figure 13.

The individual position of a vertex on a single triangle can then be reflected around the sphere to locate all the vertices of the polyhedron concerned. In general the edges of any polyhedron will consist of the lines from each vertex position that are both perpendicular to and reflected in the edges of the kaleidoscope triangle:
x| y z will thus have a single such generating line from x reflected in the line y-z;
x y |z will have two generating lines from a point on x-y that bisects z, equidistant from and reflected in x-z and y-z;
x y z| will have three generating lines from the centre of the triangle reflected in each of x-y, x-z and y-z.

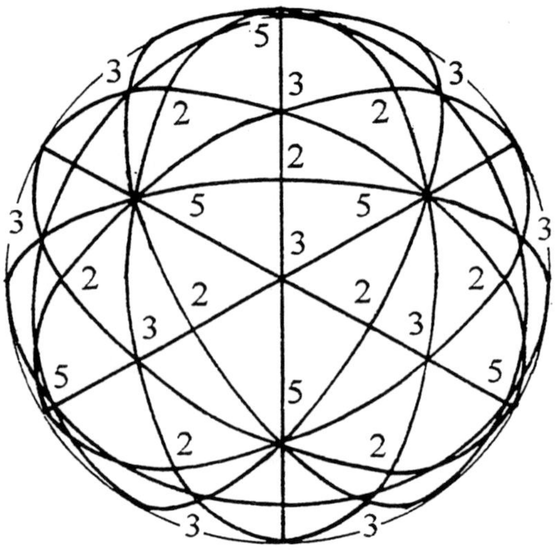

(3 2 5) Spherical Kaleidoscope

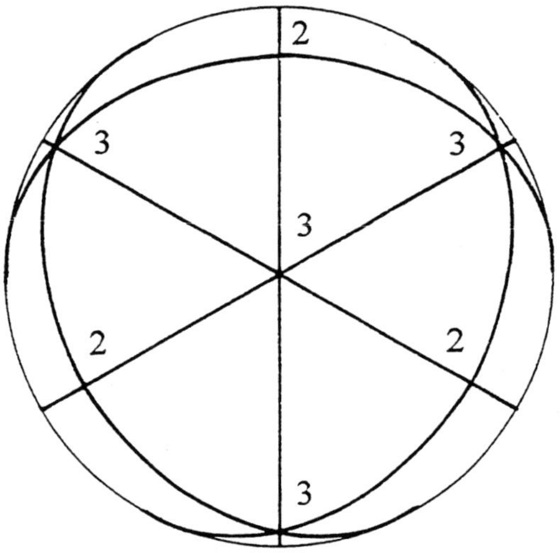

(3 2 3) Spherical Kaleidoscope

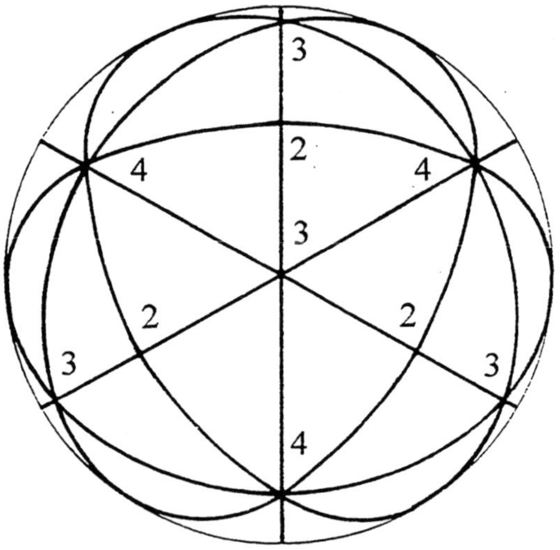

(3 2 4) Spherical Kaleidoscope

Fig.13: Spherical Kaleidoscopes

QUASITRUNCATION GENERALLY

The uniform polyhedra we have so far examined have mostly been obtained by truncating to form new faces at each vertex. As we do this the original polygon faces at a vertex become first of all their truncations, with twice as many edges, and then in turn their duals with the same number of edges as the original. The symmetry families we have been looking at have thus involved the polygons {5}, {4} and {3} along with their truncations {10}, {8} and {6}.

A polygon can also be quasitruncated by treating it as its inverse and then truncating, this process leading to both a quasitruncated polygon and then the original polygon's dual once again. The quasitruncations of the polygons {5}, {4} and {3} are truncations of {5/4}, {4/3} and {3/2} respectively, i.e. the conjugate polygons $\{^{2}5/2\}$, {8/3} and $\{^{2}3/1\}$ as shown in Figure 14. These have densities of 4, 3 and 2 respectively and comprise a double star, a star and a double polygon in turn.

Polyhedra can thus be quasitruncated by substituting the quasitruncated polygon for the normal truncated one. We can thus assemble a number of new polyhedra with appropriate vertex figures consisting of normal truncating faces and quasitruncated faces in lieu of the truncated.

As we saw with the tilings, where a double polygon occurs in a triangular vertex figure, this can be regarded as a single polygon occurring twice there, leading to a double triangular vertex figure with six edges, closely related to a 'rhombi' form. Similarly where two double polygons occur in a triangular vertex figure a more complex vertex figure can be obtained, as we shall see a little later.

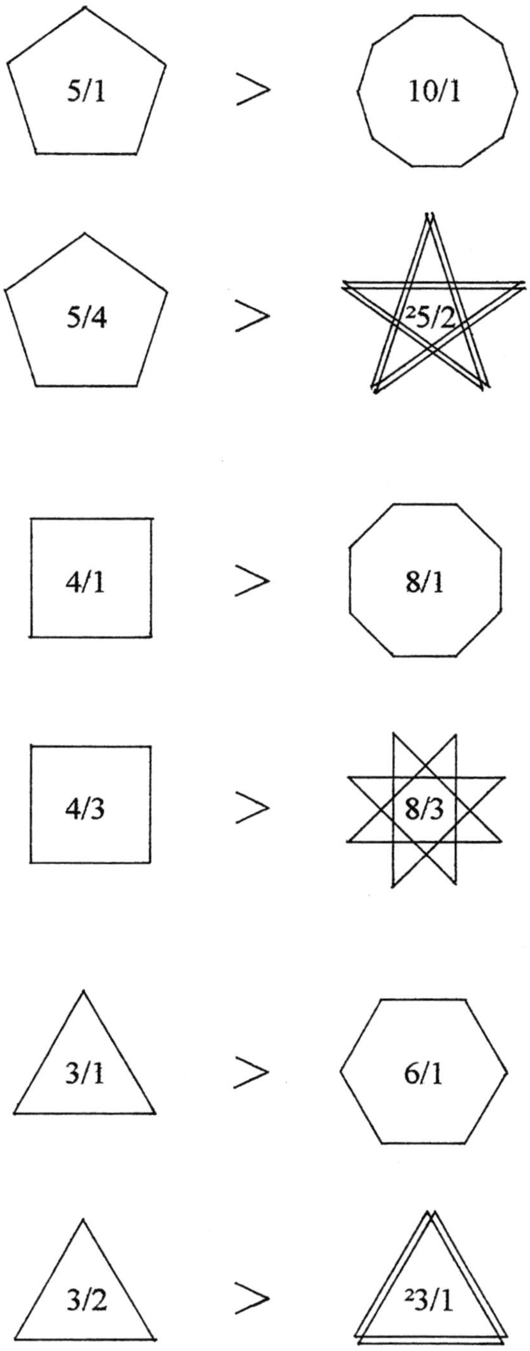

Fig. 14: Quasitruncating Polygons

QUASITRUNCATED POLYHEDRA

The quasitruncated versions of our original regular polyhedra will thus have one truncating face, the vertex figure of the original, and two quasitruncated faces at each vertex. These are shown in Figure 15 opposite and described in Table 7 below.

With any of these, if we ascribe the larger truncating polygon with positive density and the smaller valued truncated polygons with negative density we can maintain an overall density for each quasitruncated polyhedron of 1. This can best be seen with the quasitruncated cube, where to reach the centre via one of the 'crowns', we have to pass through four positive triangles and then one negative octagram (density -3).

Alternatively with the quasitruncated icosahedron, we have to pass through one negative double triangle (density -2) and then three positive internal pentagons.

These polyhedra are our first examples of non-convex star polyhedra, where faces cut through each other, much as do the edges of a star polygon. The inclusion of double polygons leads to the production of the three, four or five-fold vertex figures shown, along with three-fold edges shared by six faces (see later section on double polygon vertices). The total numbers of vertices, edges and faces can thus be made to match those of the ordinary truncations and thus still obey Euler's formula, V+F=E+2.

Symbol		V	E	F
3 2 |5/4 {3,5/2} + 2{5/2,5} Quasitruncated Dodecahedron	t{5',3}	12x5	30x3	12 20 △
3 2 |4/3 66 (92) Quasitruncated Cube	t{4',3}	24	36	6 8 △
3 2 |3/2 {3,3/2} + 2{3,3} Quasitruncated Tetrahedron	t{3',3}	4x3	6x3	4 4 △
4 2 |3/2 {4,4/2} + 2{3,4} Quasitruncated Octahedron	t{3',4}	6x4	12x3	8 6 □
5 2 |3/2 {5,5/2} + 2{3,5} Quasitruncated Icosahedron	t{3',5}	12x5	30x3	20 12 ⬠

Table 7: Quasitruncated Polyhedra

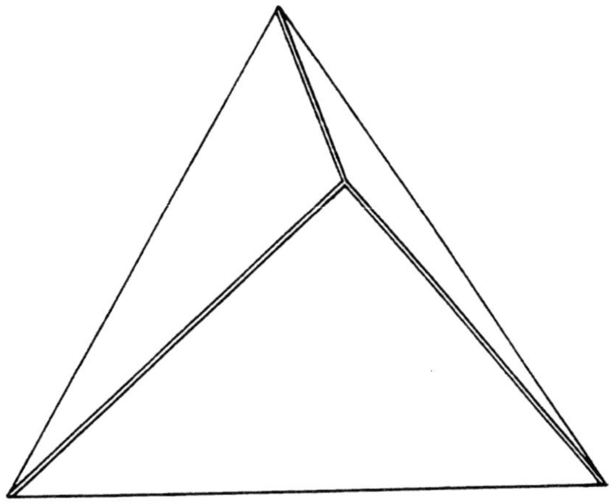

3 2 |3/2 t{3',3}

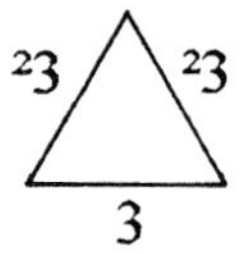

{3,3/2} + 2{3,3}
Quasitruncated Tetrahedron

Fig.15: Quasitruncated Polyhedra

3 2 |4/3 t{4',3}

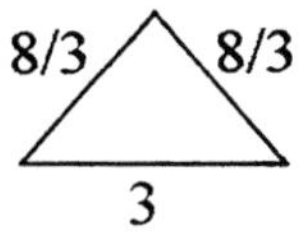

66 (92)
Quasitruncated Cube

3 2 |5/4 t{5',3}

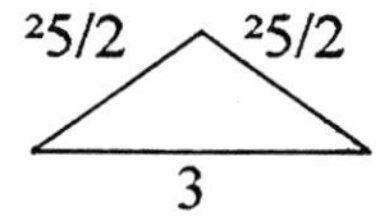

{3,5/2} + 2{5/2,5}
Quasitruncated Dodecahedron

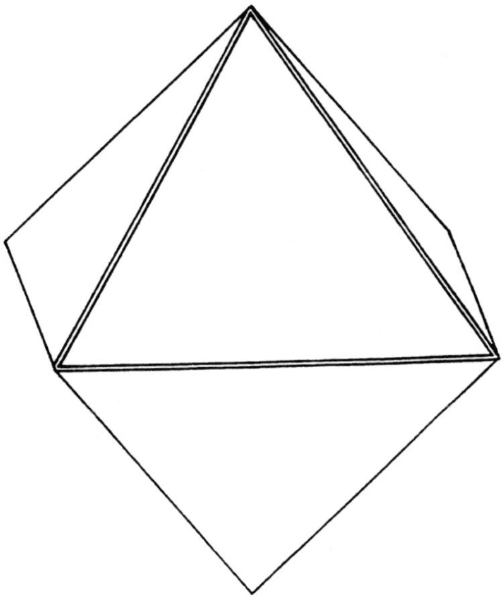

4 2 |3/2 t{3',4}

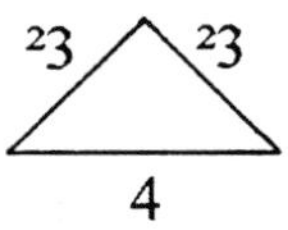

{4,4/2} + 2{3,4}
Quasitruncated Octahedron

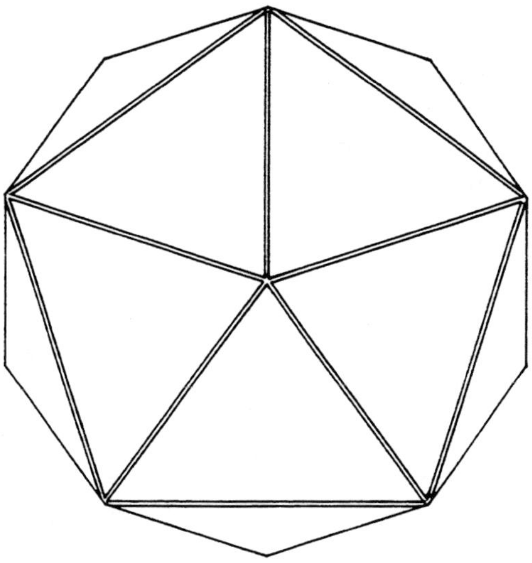

5 2 |3/2 t{3',5}

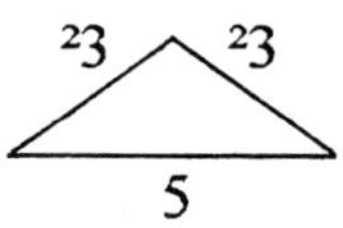

{5,5/2} + 2{3,5}
Quasitruncated Icosahedron

QUASI EVEN FACED POLYHEDRA

Continuing the quasitruncation process beyond a quasitruncated polyhedron leads back to a quasi-regular polyhedron already described. In this instance however we can regard the quasi-regular as having a rectangular vertex figure with two normal polygons opposing each other one way and two inverse polygons the other, i.e. x,y',x,y' or x',y,x',y as we read around the vertex.

Truncating these vertices in their turn and normalising the truncating faces back to squares as before, we can obtain the quasitruncated even faced polyhedra shown in Figure 16 and described in Table 8. In general these have triangular vertex figures consisting of one square truncating face, one truncated face and one quasitruncated face.

These last are often double polygon faces, which do still have an even number of edges. By attaching the prefix tri-, tetra- or penta- to their names we can distinguish between them, identifying the quasitruncated face.

As with the quasitruncated polyhedra on the previous page, only one of these is generally acknowledged as a uniform polyhedron, that given a number by Coxeter and Wenninger, the rest being thought of as 'monsters'.

Whilst the former are literally compounds of other polyhedra, the latter shown here are definitely not compounds. They are all quite easily constructed and only require the inclusion of double polygons as faces in their own right to make them fit our pattern.

Symbol		V	E	F
3 2 5/4|	t{5' / 3}	60x2	60x2	12 ✯
[2.4d]			60x1	20 ⬡
Pentaquasitr. Icosidodecahedron				30 □
3 2 4/3|	t{4' / 3}	48	72	6 ✶
67 (93)				8 ⬡
(Tetra)quasitr. Cuboctahedron				12 □
3/2 2 3|	t{3' / 3}	12x2	12x2	4 ◬
[2.4a]			12x1	4 ⬡
Triquasitruncated Octahedron				6 □
3/2 2 4|	t{3' / 4}	24x2	24x2	8 ◬
[2.4b]			24x1	6 ○
Triquasitruncated Cuboctahedron				12 □
3/2 2 5|	t{3' / 5}	60x2	60x2	20 ◬
[2.4e]			60x1	12 ○
Triquasitr. Icosidodecahedron				30 □

Table 8: Quasi Even Faced Polyhedra

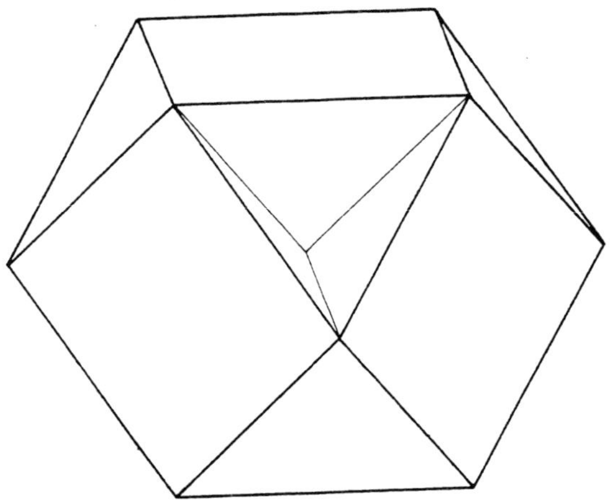

3/2 2 3| t{3' / 3}

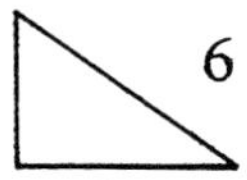

[2.4a]
Triquasitruncated Octahedron

Fig. 16: Quasi Even Faced Polyhedra

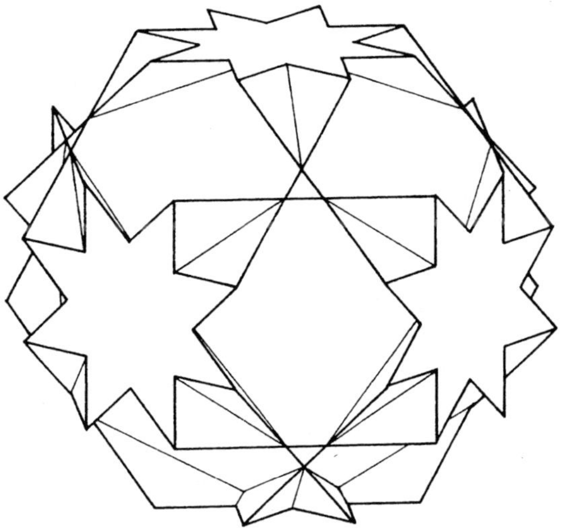

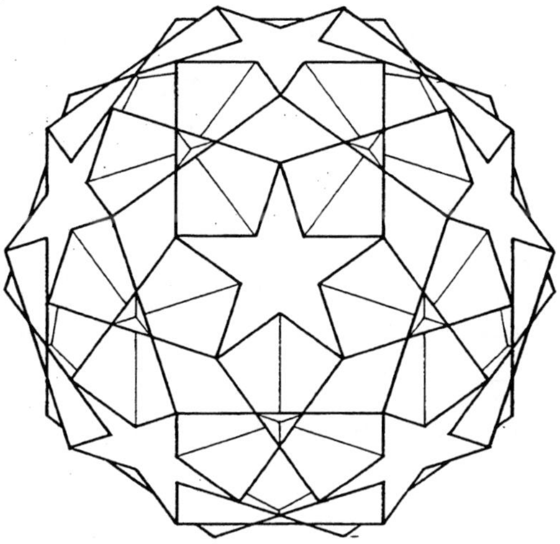

3 2 4/3| $t\begin{Bmatrix}4'\\3\end{Bmatrix}$

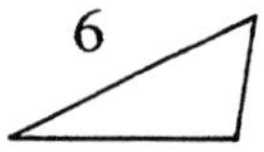

67 (93)
(Tetra)quasitruncated Cuboctahedron

3 2 5/4| $t\begin{Bmatrix}5'\\3\end{Bmatrix}$

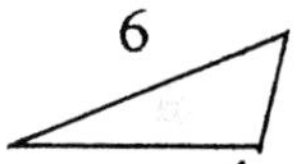

[2.4d]
Pentaquasitruncated Icosidodecahedron

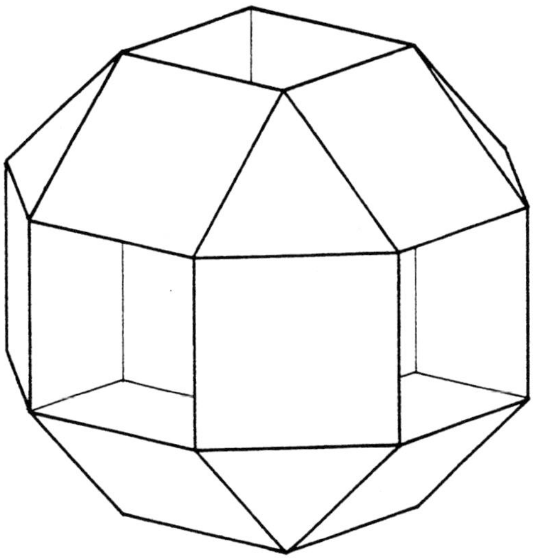

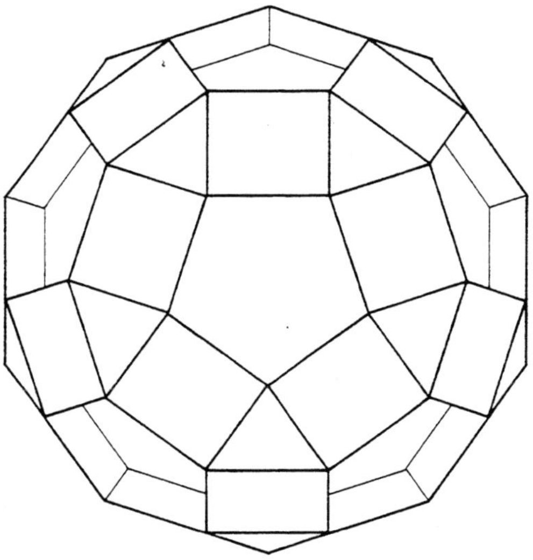

3/2 2 4| $t\begin{Bmatrix}3'\\4\end{Bmatrix}$

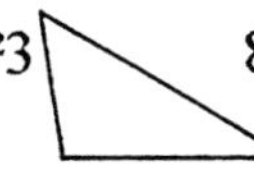

[2.4b]
Triquasitruncated Cuboctahedron

3/2 2 5| $t\begin{Bmatrix}3'\\5\end{Bmatrix}$

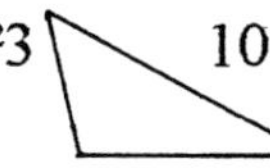

[2.4e]
Triquasitruncated Icosidodecahedron

QUASIQUASITRUNCATED POLYHEDRA

If we now consider a quasi-regular vertex figure consisting of two pairs of inverse polygons, x',y',x',y', we can produce what is in effect a double quasitruncated even faced polyhedron for each symmetry family.

This will again have a triangular vertex figure, but consisting of a square and two quasitruncated polygons, typically {4}, t{x'} and t{y'}. These quasiquasitruncated polyhedra are shown in Figure 17 and described in Table 9 that follows.

With both these and the quasi even faced polyhedra already described, if we make the largest polygon at each vertex positive and the other two negative, we can maintain the density at 1.

3/2 2 5/4| t$\begin{Bmatrix}3'\\5'\end{Bmatrix}$

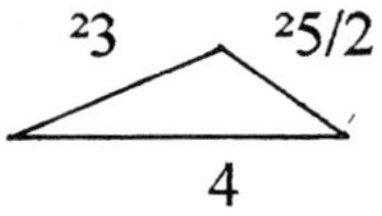

[2.4f]
Quasiquasitruncated Icosidodecahedron

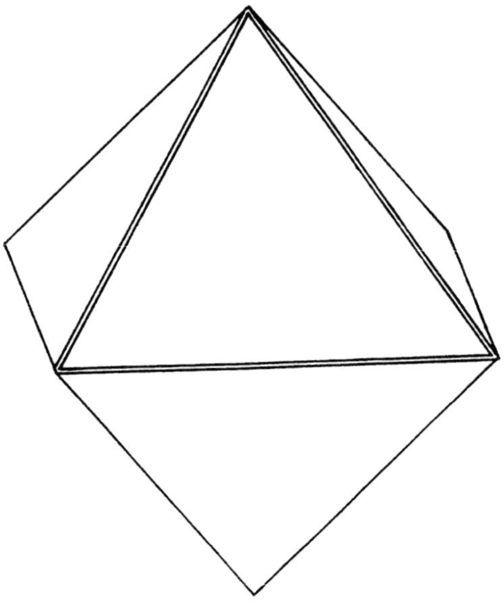

3/2 2 3/2| t$\begin{Bmatrix}3'\\3'\end{Bmatrix}$

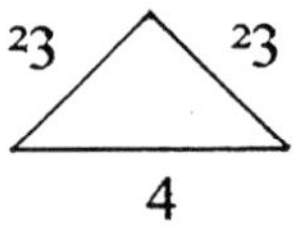

{4,4/2} + 2{3,4}
Quasiquasitruncated Octahedron

3/2 2 4/3| t$\begin{Bmatrix}3'\\4'\end{Bmatrix}$

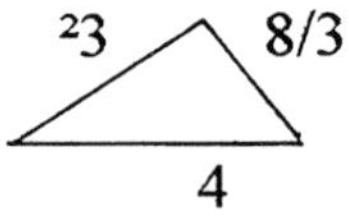

[2.4c]
Quasiquasitruncated Cuboctahedron

Fig.17: Quasiquasitruncated Polyhedra

QUASIRHOMBIHEDRA

Continuing the quasitruncation of a quasiquasitruncated polyhedron we arrive back at the original rhombihedron, but with two squares and two inverse polygons opposite each other around a vertex.

However, doing the same with either of the singly quasitruncated even faced polyhedra leads to a quasirhombihedron for each symmetry family. These have two squares plus one normal and one inverse polygon opposing each other at each vertex.

If the inverse is then removed, changing its direction by twisting the vertex figure, the square faces cross over each other. These quasirhombihedra are shown in Figure 18 and described overleaf in Table 10. Two of these have multiple vertices as shown in Figure 20 that follows.

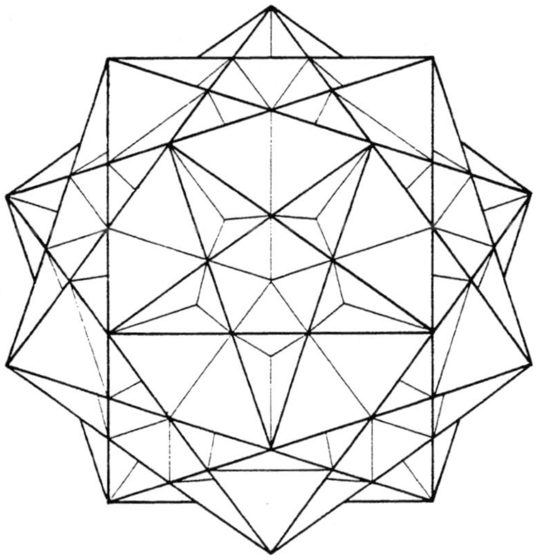

3/2 5 |2   r' $\begin{Bmatrix}3\\5\end{Bmatrix}$

[2.4h]

Quasirhombicosidodecahedron

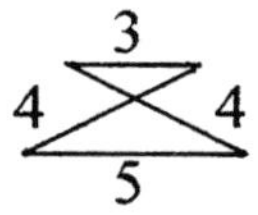

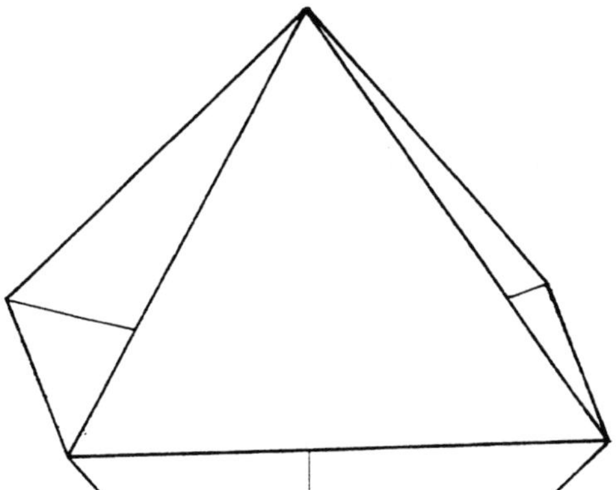

3/2 3 |2   r' $\begin{Bmatrix}3\\3\end{Bmatrix}$

36 (67)

Quasicuboctahedron

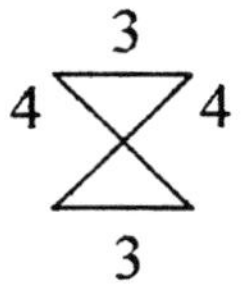

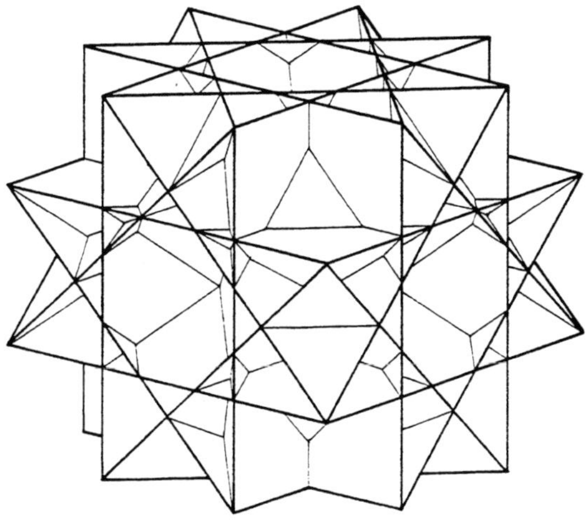

3/2 4 |2   r' $\begin{Bmatrix}3\\4\end{Bmatrix}$

59 (85)

Quasirhombicuboctahedron

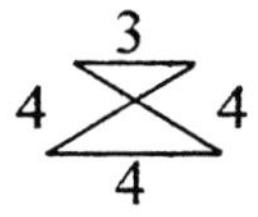

Fig. 18: Quasirhombihedra

QUASISNUB POLYHEDRA

The quasirhombihedra, like the rhombihedra, each have two square faces per vertex, which can be omitted in a similar way. This allows a twist to be given with two triangles taking each square's place, thus producing a quasisnub polyhedron. These are shown in Figure 19 and described in Table 11, appearing to all intents and purposes to be simply a tetrahedron, an octahedron and an icosahedron.

Like the quasitruncated polyhedra seen earlier, they do however have additional faces internally: triangles, squares and pentagons respectively. These make up three, four and five-fold vertices along with five-fold edges as shown in the vertex figures of Figure 20 opposite. The numbers of vertices and edges thus match the snub polyhedra and obey Euler's formula.

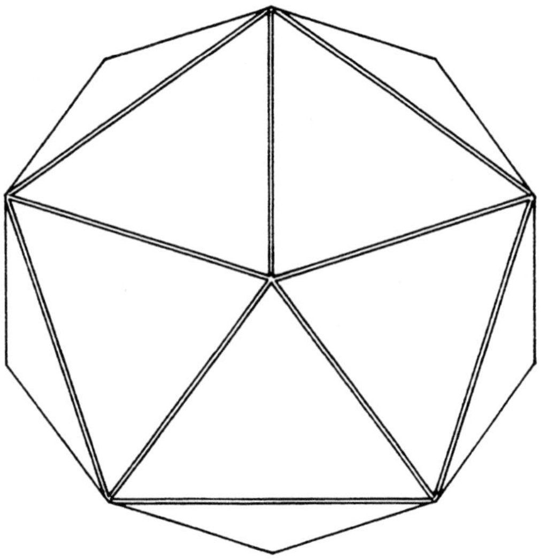

|3/2 2 5 $s'\begin{Bmatrix}3\\5\end{Bmatrix}$

{5,5/2} + 4{3,5}

Quasisnub Dodecahedron

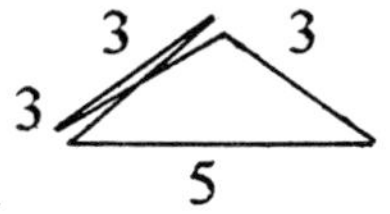

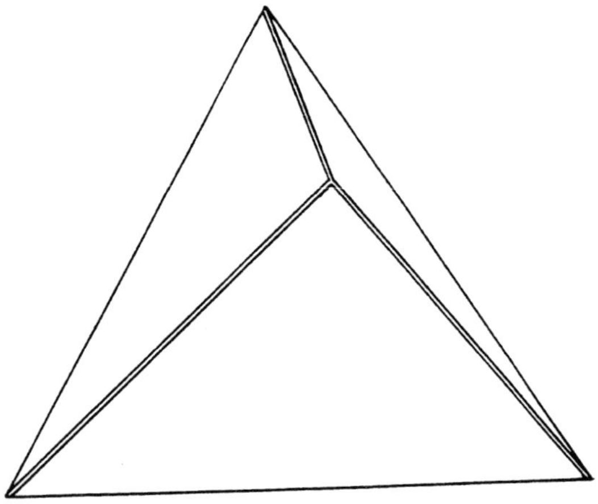

|3/2 2 3 $s'\begin{Bmatrix}3\\3\end{Bmatrix}$

{3,3/2} + 4{3,3}

Quasisnub Tetrahedron

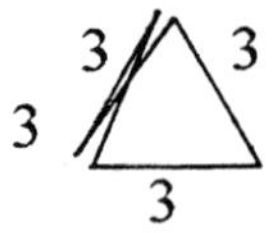

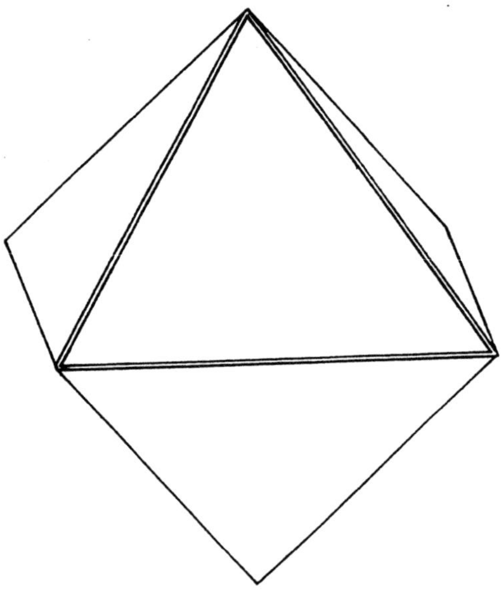

|3/2 2 4 $s'\begin{Bmatrix}3\\4\end{Bmatrix}$

{4,4/2} + 4{3,4}

Quasisnub Cube

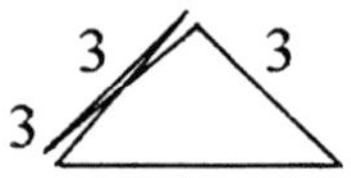

Fig.19: Quasisnub Polyhedra

Symbol		V	E	F
3/2 2 5/4|	t{3'	20x6	60x3	12 ☆
[2.4f]	5'}			20 △
Quasiquasitr. Icosidodecahedron				30 □
3/2 2 4/3|	t{3'	24x2	24x2	6 ☆
[2.4c]	4'}		24x1	8 △
Quasiquasitr. Cuboctahedron				12 □
3/2 2 3/2|	t{3'	6x4	12x3	4 △
t{3',4}	3'}			4 △
Quasiquasitruncated Octahedron				6 □

Table 9: Quasiquasitruncated Polyhedra

3/2 5 |2	r'{3	20x3	60x2	12 ⬠
[2.4h]	5}			20 △
Quasirhombicosidodecahedron				30 □
3/2 4 |2	r'{3	24	48	6 □
59 (85)	4}			8 △
Quasirhombicuboctahedron				12 □
3/2 3 |2	r'{3	6x2	12x2	4 △
36 (67)	3}			4 △
Quasicuboctahedron				6 □

Table 10: Quasirhombihedra

|3/2 2 5	s'{3	12x5	30x5	12 ⬠
	5}			20 △
Quasisnub Dodecahedron				60 △
|3/2 2 4	s'{3	6x4	12x5	6 □
	4}			8 △
Quasisnub Cube				24 △
|3/2 2 3	s'{3	4x3	6x5	4 △
	3}			4 △
Quasisnub Tetrahedron				12 △

Table 11: Quasisnub Polyhedra

POLYHEDRAL SHELLS

Comparing the quasirhombihedra with the quasiquasitruncated polyhedra on the facing page we can observe certain similarities of form. Each pair can be said to share the same polyhedral shell, a property also to be seen with the several others amongst our quasitruncated forms.

What is happening here is that we are discovering the inscribed polygons that lurk within any vertex figure greater than a triangle. The next section will look a little more closely at this phenomenon with particular emphasis on the rhombihedra. Although this generally involves multiple vertices and edges and often double polygon faces, the results obtained will hopefully justify their inclusion here.

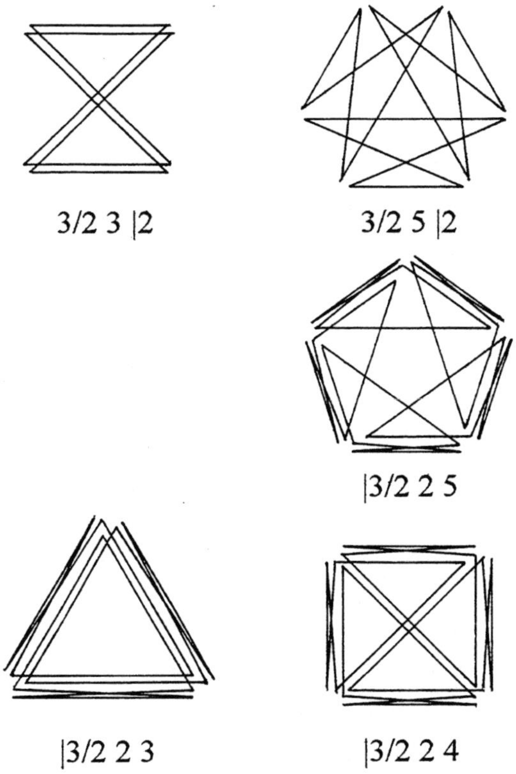

Fig.20: Multiple Vertex Figures

P-TYPE RHOMBIHEDRA

The simplest vertex figures greater than a triangle are quadrilaterals which, if we include the diagonals, have six possible faces present. Our original rhombihedra miss these diagonals out, but by including the diagonals and then omitting other faces we can produce new varieties of polyhedra.

If we omit the perimeter square faces of a rhombihedron in this way we get a vertex figure reminiscent of our quasirhombihedra, consisting of two parent faces from the particular symmetry family concerned and the two diagonal faces crossed. These are the p-type rhombihedra, which by analogy with the quasirhombihedra can be assigned Coxeter notations as shown in Figure 21. These notations indicate the possible existence of further symmetry families.....

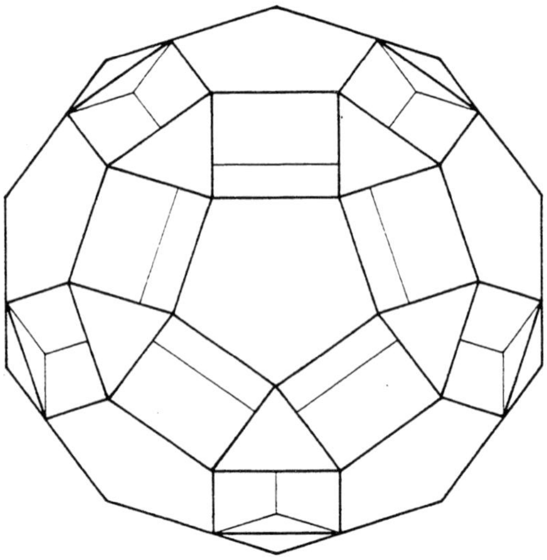

3 5/4 |5 $p\left\{\begin{matrix}3\\5\end{matrix}\right\}$
3/2 5 |5
42 (72)
Small Dodecicosidodecahedron

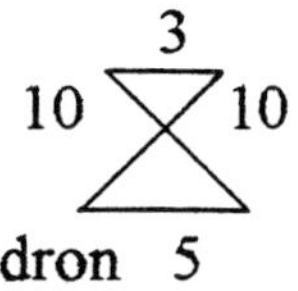

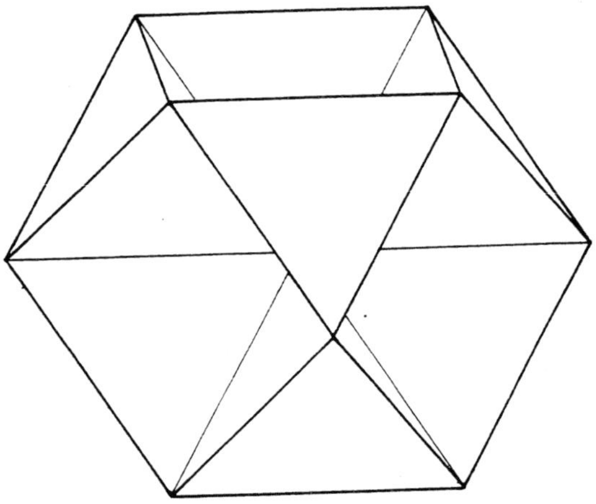

3 3/2 |3 $p\left\{\begin{matrix}3\\3\end{matrix}\right\}$
3/2 3 |3
37 (68)
Octahemioctahedron

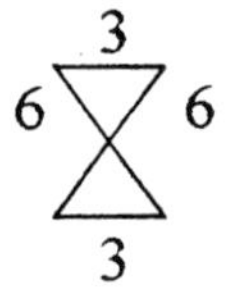

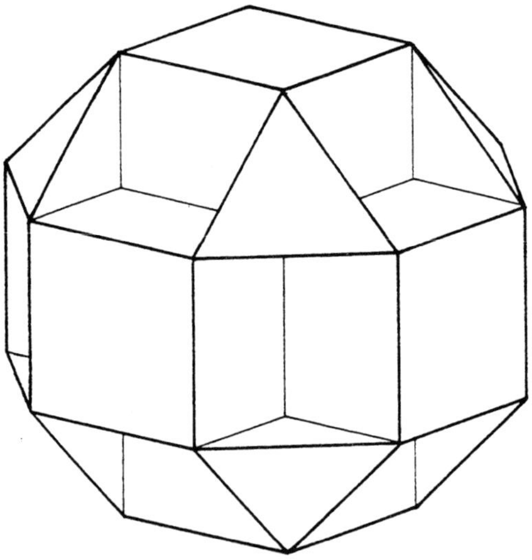

3 4/3 |4 $p\left\{\begin{matrix}3\\4\end{matrix}\right\}$
3/2 4 |4
38 (69)
Small Cubicuboctahedron

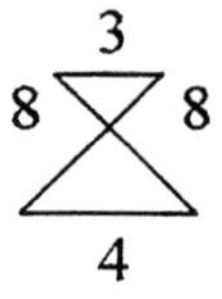

Fig.21: P-type Rhombihedra

Q-TYPE RHOMBIHEDRA

A third alternative vertex figure can be derived from the original rhombihedra by adding the diagonals and then omitting the parent faces. These q-type rhombihedra are shown in Figure 22. They consist of square and diagonal faces only, which give rise to the first two figures in their symbols, the paired figures at the end coming from the missing parent faces.

These p- and q-types share the same shells as the original rhombihedra, as do the triquasitruncated polyhedra of Figure 16. These last have only one of the parent faces, that other than a triangle, missing. We will see in the next volume the other possibility, that with only the parent triangle faces missing, the quasitruncated even faced polyhedra with symbols 4 2 4/2| and 5 2 5/2|, a further hint of things to come!

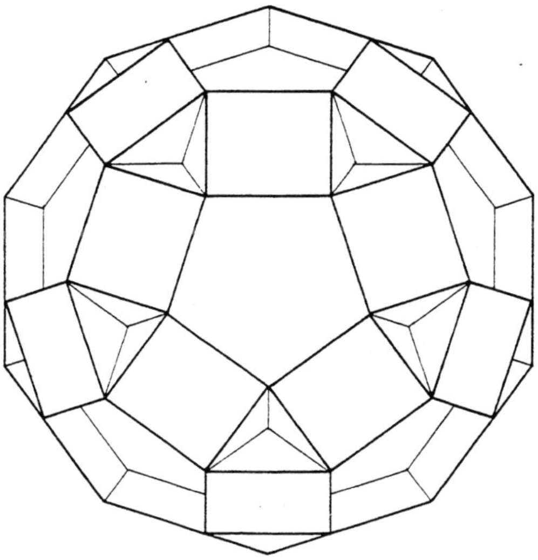

5 2 3/2| q{3}
5/2| {5}

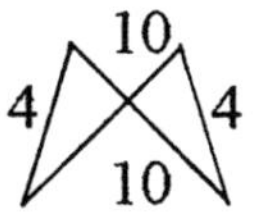

46 (74)
Small Rhombidodecahedron

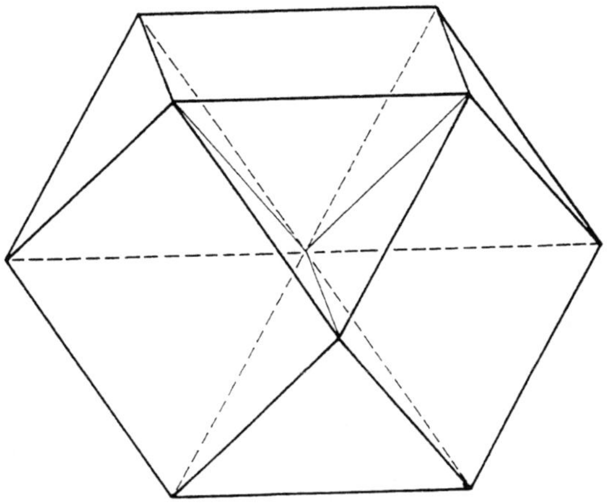

3 2 3/2| q{3}
3/2| {3}

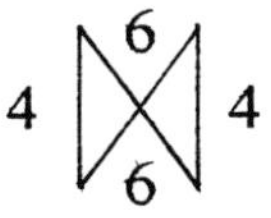

51 (78)
Cubohemioctahedron

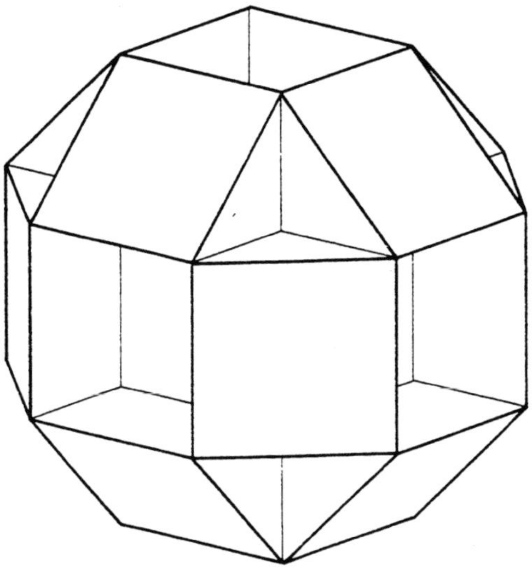

4 2 3/2| q{3}
4/2| {4}

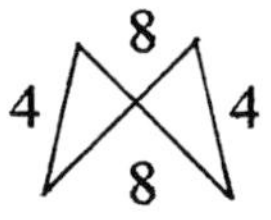

60 (86)
Small Rhombihexahedron

Fig.22: Q-type Rhombihedra

P'-TYPE QUASIRHOMBIHEDRA

If we attempt the same procedure with the quasirhombihedra, first of all we have to include back in the missing perimeter faces of the vertex figure.

These turn out to be the conjugate polygons to the missing diagonal faces of the rhombihedra, and accordingly comprise one double polygon $\{^{2}3/1\}$, one star polygon $\{8/3\}$ and one double star polygon $\{^{2}5/2\}$.

We can then omit the diagonal square faces of the quasirhombihedra to obtain the rhombihedron-like quadrilateral vertex figures of the p'-type quasirhombihedra as shown in Figure 23.

These interestingly seem to indicate the existence of the same new symmetry families as did the p-type rhombihedra.

3 5 |5/4 $p'\begin{Bmatrix}3\\5\end{Bmatrix}$

Vertex figure: top 3, sides $^{2}5/2$ and $^{2}5/2$, base 5

Small Quasidodecicosidodecahedron

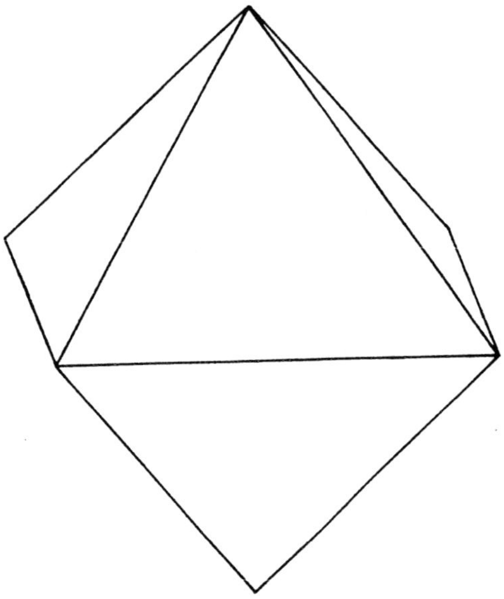

3 3 |3/2 $p'\begin{Bmatrix}3\\3\end{Bmatrix}$

Vertex figure: top 3, sides $^{2}3$ and $^{2}3$, base 3

17 (2)

Double Octahedron

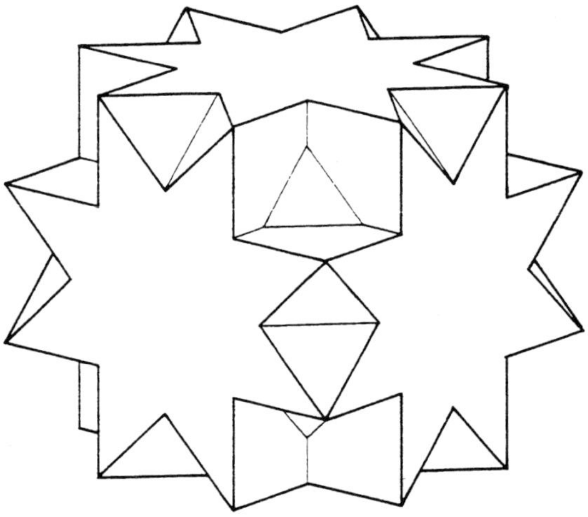

3 4 |4/3 $p'\begin{Bmatrix}3\\4\end{Bmatrix}$

Vertex figure: top 3, sides 8/3 and 8/3, base 4

50 (77)

Great Cubicuboctahedron

Fig.23: P'-type Quasirhombihedra

Q'-TYPE QUASIRHOMBIHEDRA

Obviously taking this process to its logical conclusion, we can also obtain a third alternative q'-type quadrilateral vertex figure composed of only the diagonal square faces and the inserted conjugates. These are shown in Figure 24.

The basic outer shell in this instance is the p'-type with the quasirhombihedron itself and the q'-type filling in the diagonals in different ways.

These three also share their shells with another type we have already seen, the quasiquasitruncated polyhedra. The latter have only one of the parent faces missing and are again shadowed by alternatives with the other (triangular) parent face missing, 4/3 2 4/2| and 5/4 2 5/2|, which we shall again see in the next volume.

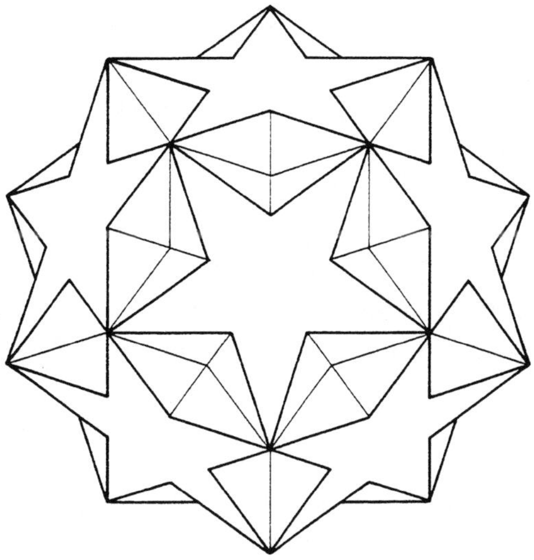

5/4 2 3/2| q'{3}
5/2| {5}

4
^{2}5/2 ^{2}5/2
4

Small Quasirhombidodecahedron

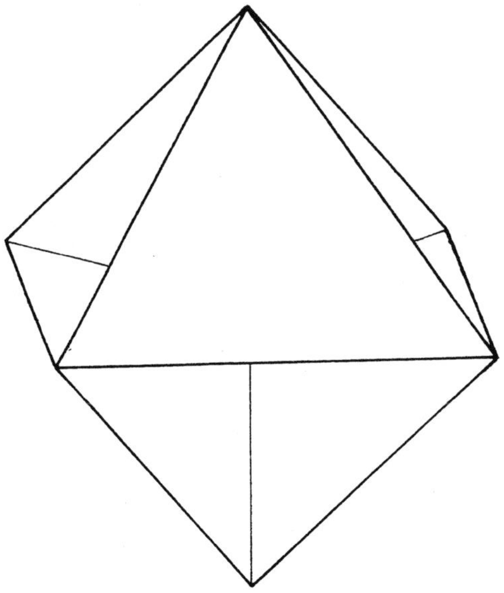

3/2 2 3/2| q'{3}
3/2| {3}
36 (67)

4
^{2}3 ^{2}3
4

Double Tetrahemihexahedron

4/3 2 3/2| q'{3}
4/2| {4}
82 (103)

4
8/3 8/3
4

Great Rhombihexahedron

Fig.24: Q'-type Quasirhombihedra

DOUBLE POLYGON VERTICES

The similarities and sharing of basic polyhedral shells between the quasitruncated even faced polyhedra and the rhombihedra can be simply explained in terms of double polygons.

If we treat the double polygons occuring in the vertices of the quasitruncated even faced polyhedra as single polygons occuring there twice, we can effectively visit each vertex twice by making two interlocking triangles as shown in Figure 25.

What we then find is that these double triangular vertex figures match exactly the corresponding quadrilateral vertex figures of the rhombihedra. The polyhedra produced thus appear like rhombihedra with one type of parent face missing.

We have already noted earlier when discussing q-types and the triquasitruncated polyhedra, that in order to produce a similar vertex figure with the other parent face missing, we need to invent both {4,4/2} and {5,5/2} symmetry families.

In fact the fourth example shown, 3 2 5/4|, the pentaquasitruncated icosidodecahedron, would appear to fit in with there being a rhombihedron in the {5,5/2} family since the missing face needed to complete the quadrilateral is a pentagon. Turning this one on its head we would need to start with a 3 2 5/2| to obtain the same vertex figure but with a double pentagon added and the double pentagram omitted: a possible {3,5/2} family member. The plot thickens!

Our final example shows that carrying out the same procedure on the one double polygon quasiquasitruncated polyhedron produces a match for the quasirhombihedron

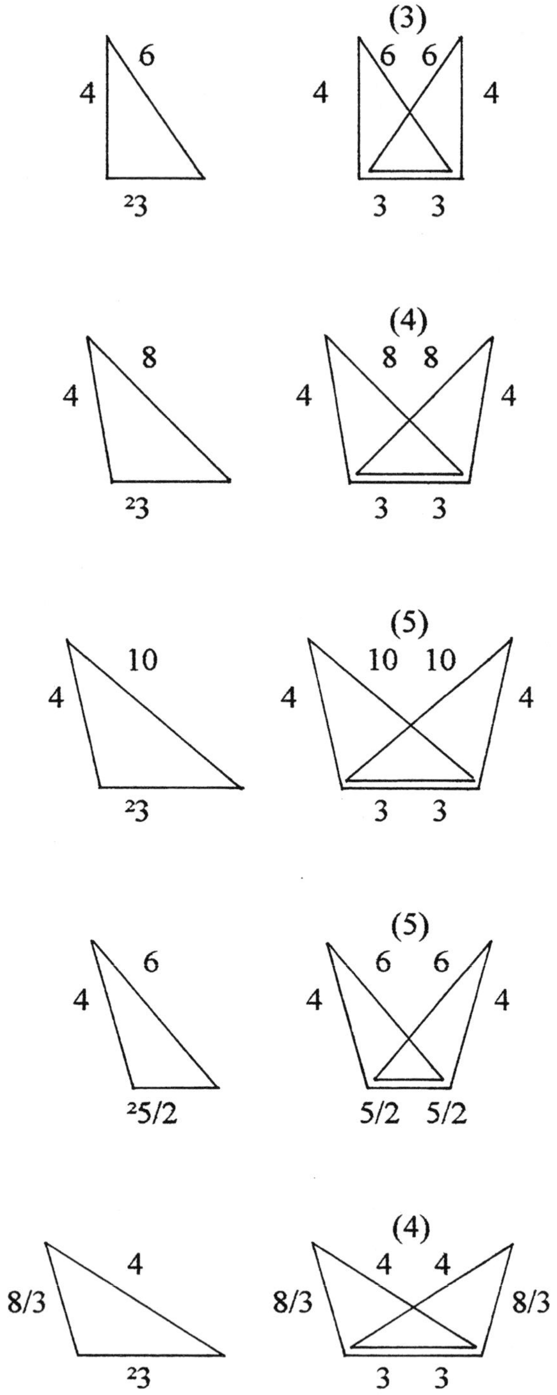

Fig.25: Double Polygon Quadrilaterals

A similar process to that just discussed can be applied to the quasitruncated polyhedra with two double polygons per vertex.

This at first produces a quadrilateral vertex figure, but with spare double polygons on either side. Repeating the process for these we can eventually close the vertex figure, which usually forms a regular inscribed polygon.

The appearance of the first three of these as seemingly regular polyhedra is thus explained, however each of them actually consists of a double outer triangular faced platonic solid with a single inscribed internal configuration. Those figures inscibed within the 'octahedron' and 'icosahedron' again suggest the existence of {4,4/2} and {5,5/2} as something we should consider. These will in fact be seen in the next volume on the star and cross polyhedra to be real members of uniform polyhedral families with triple density. Taken together with the two enveloping solids of density –1 each we still maintain a total density of 1 for each.

The fourth example shown here is again a little different. It appears to be a double {5/2,5} with an inscribed {3,5/2}. This latter will also be seen in due course with sevenfold density, offset by the two former with density –3 each, again totalling 1.

The final example shown is another quasiquasitruncation with double polygons of two different types at each vertex, forming an irregular hexagon vertex figure. The shell is a double Small Ditrigonal Icosidodecahedron (39 (70)) with a compound of five cubes inscribed within it. This vertex figure also contains pentagons deep within it and occurs in many guises.

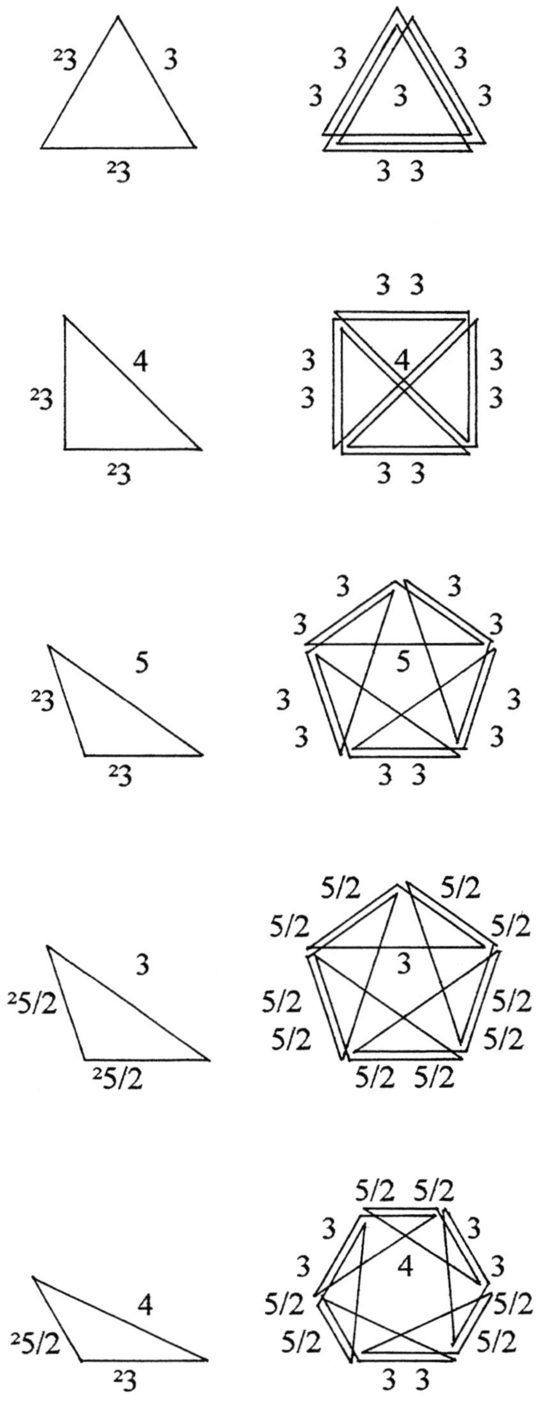

Fig.26: Double Polygon Polygons

MULTIPLE VERTEX FIGURES

Examination of the multiple vertex figures produced from double polygons can tell us much about the actual polyhedra concerned.

Since each edge of a vertex figure represents a face and each vertex of the vertex figure an edge of the polyhedron, the number of pairs of edges at each vertex of the vertex figure is a measure of how many edges of the polyhedron are coincident. Similarly for the quasi even faced, quasitruncated and quasiquasitruncated polyhedra the number of triangles making up the multiple vertex figure formed is a measure of how many vertices are coincident at any one vertex.

It is these multiples which, fed back into our tabulations, have maintained the values of V and E, so that Euler's formula continues to apply.

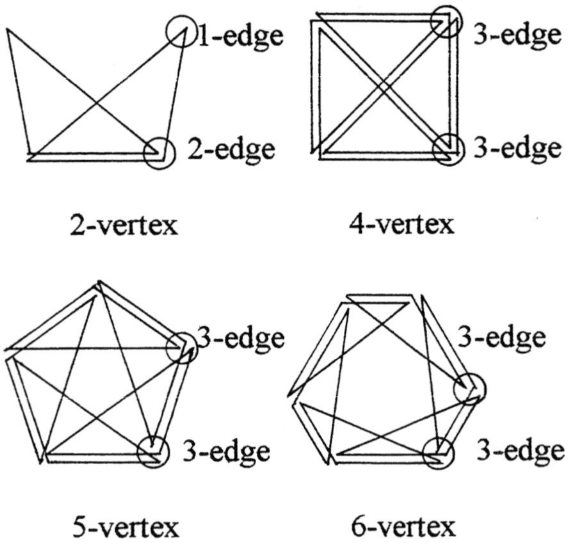

Fig.27: Multiple Edges and Vertices

THREE MORE POLYGONS

The preceding pages have given us heavy hints that the polygons {4/2} and {5/2} will be needed in due course as part of our polygonal vocabulary.

This is particularly so when we come to deal with the star and cross polyhedra in the next volume, but also true to some extent for the following section in this volume which will deal with the sub-families of the polyhedral symmetry families we have so far examined.

This is therefore a convenient point at which to introduce these new polygons. In addition to {4/2}, which is self inverse, we have only to consider {5/2} and its inverse {5/3}, along with the truncations of all three, $\{^{2}4/1\}$, $\{^{2}5/1\}$ and {10/3}, as shown below in Figure 28.

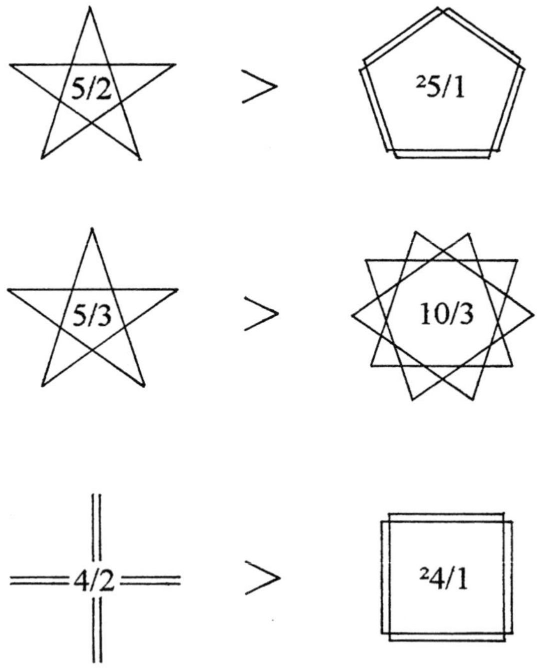

Fig.28: Truncated Polygons

VERTEX FIGURE CYCLES

Before summarising the three symmetry families {3,3}, {3,4} and {3,5} that we have been examining, it is perhaps worth looking at one other aspect of their vertex figures that leads us on towards the star and cross polyhedra.

If we take the rhombi and quasirhombihedra for each family and split apart their quadrilateral vertex figures into two constituent triangles, we can take these as even faced vertex figures from which further truncation will yield yet more rhombihedra.

The results shown in Figure 29 are particularly interesting for the octahedral and icosahedral families. These each produce two of the original rhombihedra plus two entirely new ones, again indicating the possible existence of rhombihedra for {4,4/2} and {5,5/2} families...

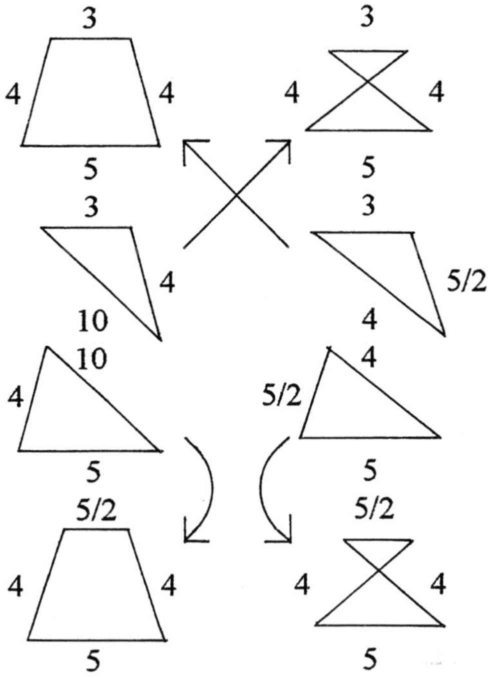

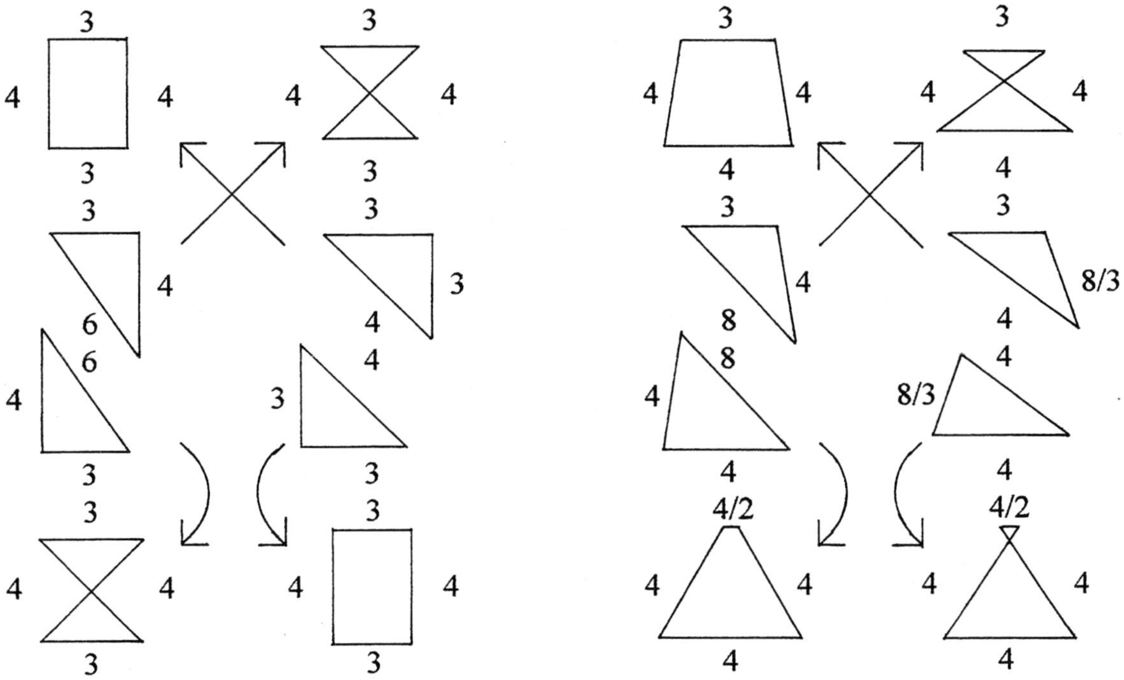

Fig.29: Rhombi Vertex Figure Cycles

THREE COMPLETE? FAMILIES

We saw on page 24 that the vertices of various regular polyhedra and their immediate truncations can be plotted in terms of a spherical kaleidoscope of Möbius triangles.

This property also holds for the various quasitruncations, but their vertex positions are determined by the extended edges and external bisectors of the Möbius triangles, falling within what are known as 'colunar' triangles.

A 'lune' consists of two particular edges of the Möbius triangle, extended as great circles until they cross again at a point diametrically opposed to the original crossing in the Möbius triangle itself.

Any Möbius triangle (x 2 y) will thus have three colunar triangles corresponding to (x' 2 y), (x 2 y') and (x' 2 y'). These are best identified by the even faced polyhedra that occur at their centres, a quasitruncation, a quasitruncation and a quasiquasitruncation respectively.

Each colunar triangle consists of a whole number of smaller Möbius triangles.

The Möbius triangles and their associated colunar triangles are shown for our three symmetry families in Figures 30, 31 and 32. Here the terms {3,3}, {3,4} and {3,5} we have used so far have been replaced by the equivalent (3 2 3), (3 2 4) and (3 2 5), which will be found more useful from here on in.

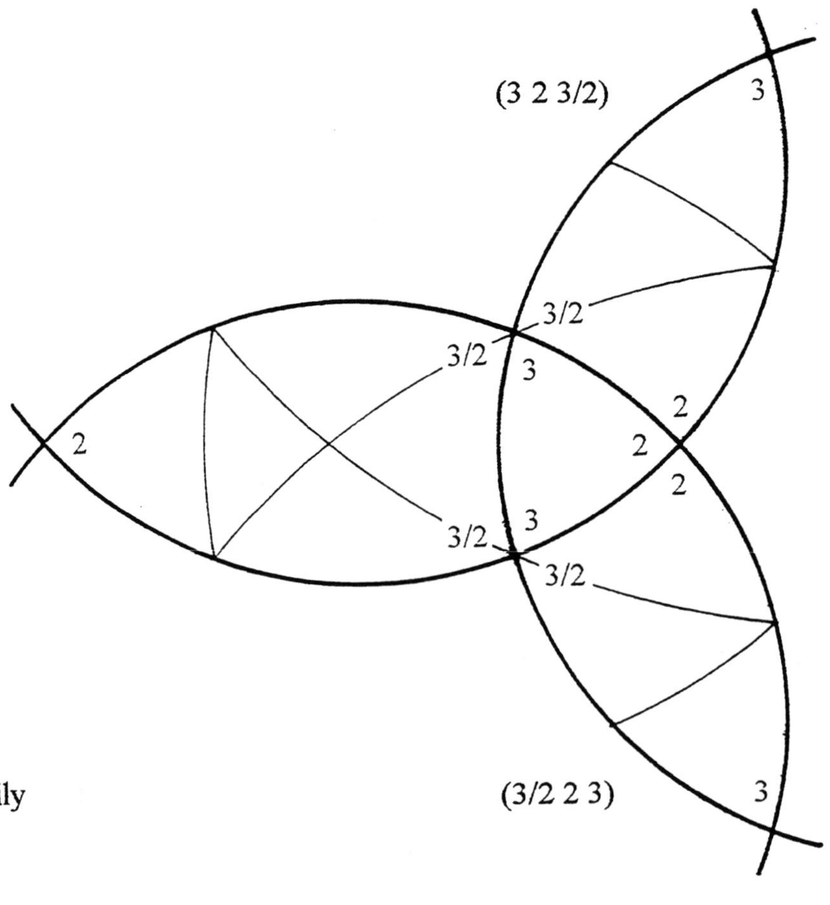

Fig.30: The (3 2 3) Family

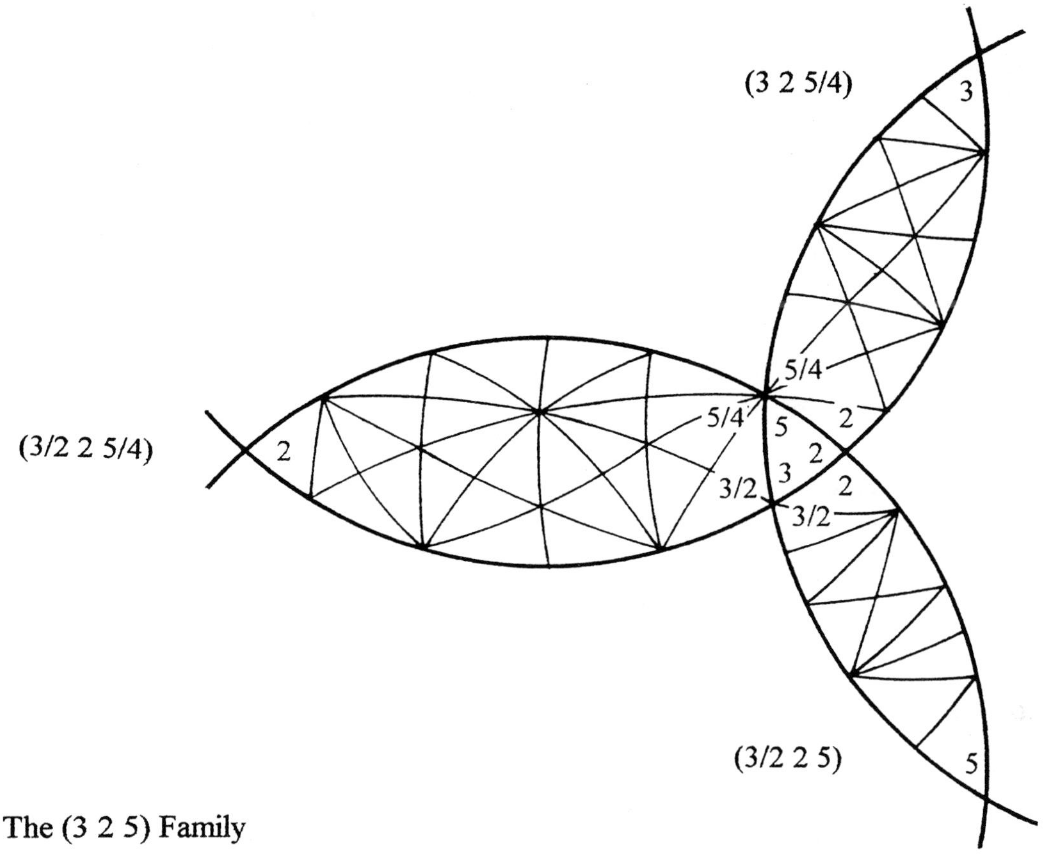

Fig.31: The (3 2 5) Family

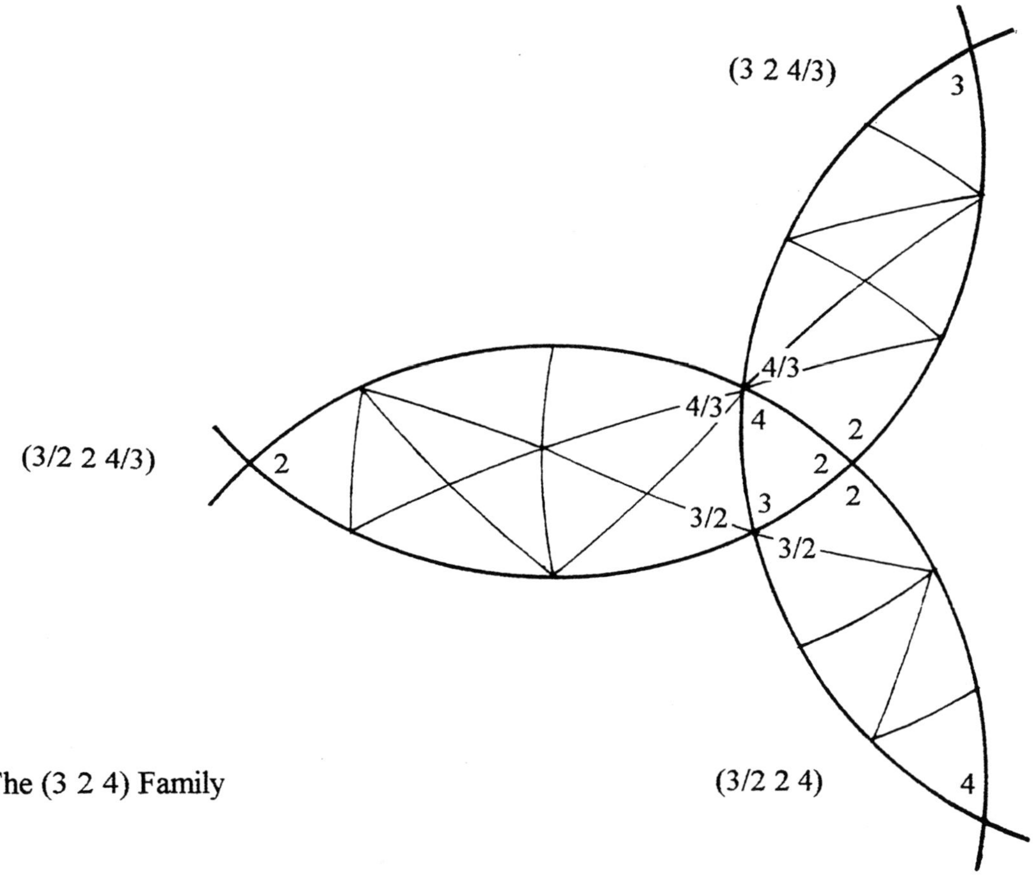

Fig.32: The (3 2 4) Family

By taking pairs of Möbius triangles back to back, we can form new larger triangles known as Schwarz triangles.

If {n} is the polygon represented by a particular vertex of a Möbius triangle, then that vertex will have an internal angle of (180/n)°. Two such vertices abutted will have an internal angle of (360/n)°, thus representing the polygon {n/2}.

The possible Schwarz triangles consisting of two Möbius triangles are thus (3 3/2 3), (3 4/2 3), (3 5/2 3), (4 3/2 4) and (5 3/2 5) as shown in Figure 33. They act similarly to their constituent Möbius triangles, also reflecting around a spherical kaleidoscope to cover the sphere and forming the basis of five sub-families related to our original three families of polyhedra.

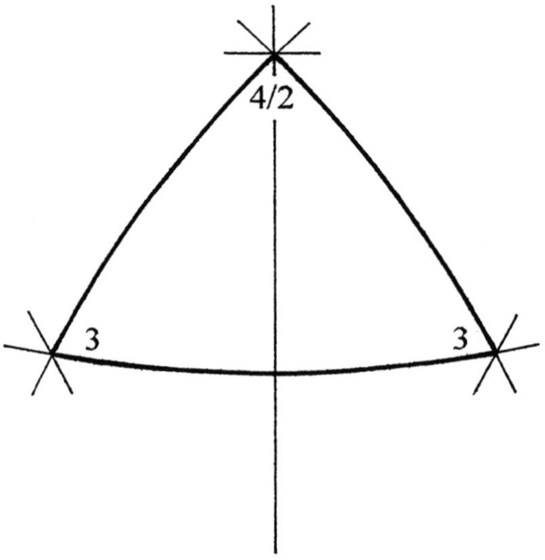

(3 4/2 3)

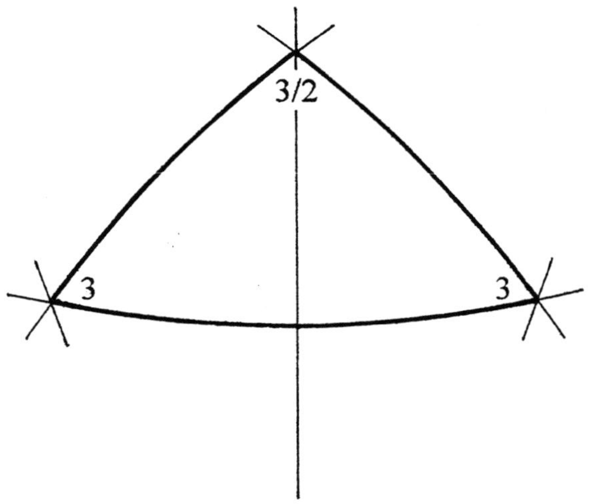

(3 3/2 3)

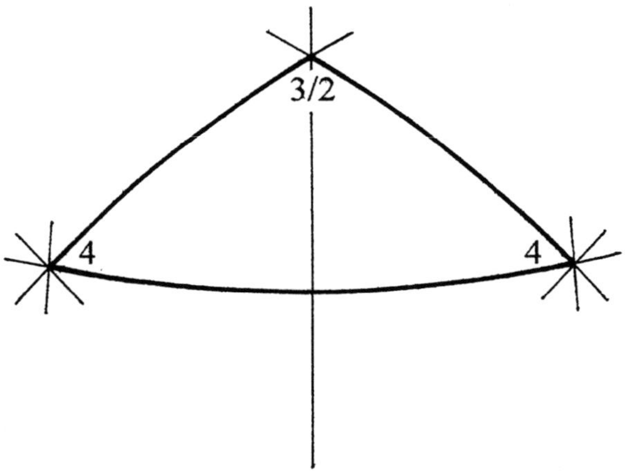

(4 3/2 4)

Fig.33: Sub-Family Schwarz Triangles

DOUBLE POLYHEDRA

These double Möbius triangle Schwarz triangles can be generalised as (x y/2 x) as shown in Figure 34 below.

At the apex of the triangle we find y/2| x x, which is a double version of the parent polyhedron y| 2 x.

In the centre we find the even faced x y/2 x|, which is a double version of the truncated parent polyhedron y 2 |x.

At the centre of the base we find x x |y/2, which is a double version of the parent quasi-regular polyhedron 2| x y.

And finally in the centre of the lower colunar triangle we find the even faced x' y/2 x'|, which is a double version of the quasitruncated parent polyhedron y 2 |x'.

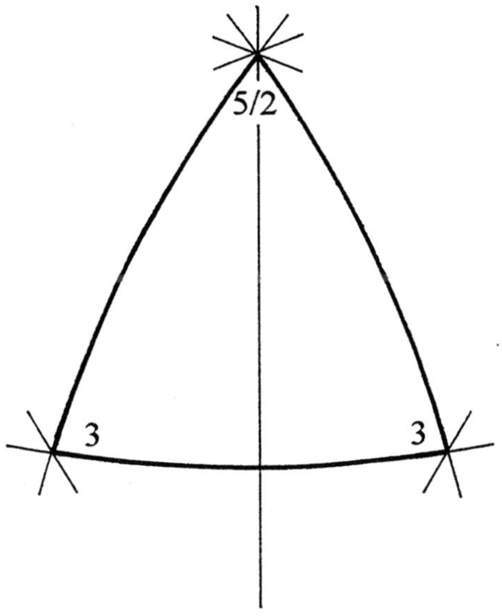

(3 5/2 3)

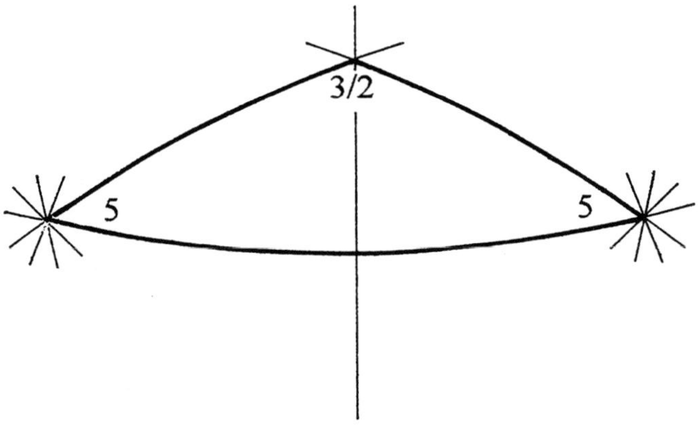

(5 3/2 5)

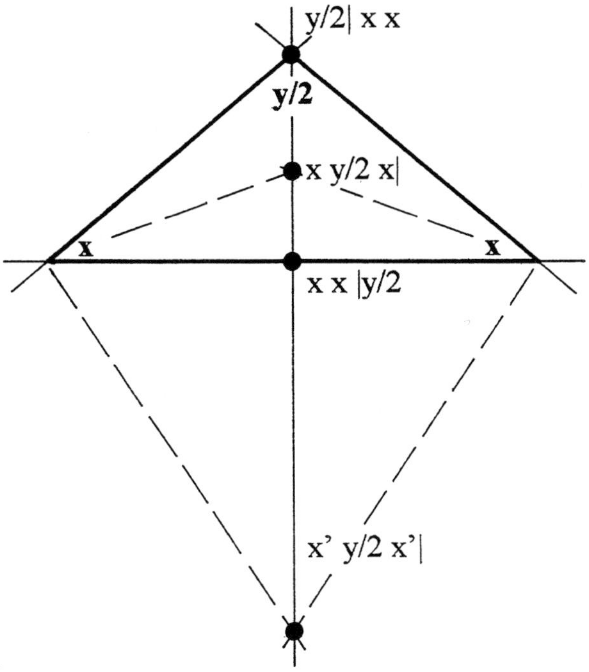

Fig.34:Typical Sub-Family Schwarz Triangle

MORE QUASI-REGULAR POLYHEDRA

The double polyhedra obviously have a density of 2, but fortunately doubles are not the sole products of these new Schwarz triangles as might be feared. Taking the lower vertex of these sub-family triangles (either will do since they are reflections of each other and equivalent), we find the polyhedron x| y/2 x. In general this has a {2x} vertex figure with alternating {x} and {y/2} faces around it.

When y is 4 or 5 we get the hexagonal equivalent of the rectangular vertex figure of a quasi-regular polyhedron. This produces two new polyhedra which we have already seen in other guises, the Stella Octangula described by Kepler and the related Small Ditrigonal Icosidodecahedron.

When y is 3, however, the inverse polygon {3/2} involved causes the vertex figure to fold up from its original hexagon, octagon or decagon into a triangle, square or pentagon respectively, as shown in Figure 35.

The results here are once again three of the original Platonic solids, but here inscribed with their inner polygons.

These polyhedra all have a density of 2 and are described in Table 12 below. At first they do not seem to obey Euler's formula, but by counting all their double density {y/2} faces twice, we can achieve the result that V+F=E+4, which matches the result we would obtain for the double polyhedra.

Symbol	V	E	F
3| 5/2 3	20	60	12 ✰
39 (70)			20 △
Small Ditrigonal Icosidodecahedron			
3| 4/2 3	8	24	6 ✕
(19) [1.4a]			8 △
Stella Octangula			
3| 3/2 3	4	6x2	4 △
{3,3/2} + {3,3}			4 △
Inscribed Tetrahedron			
4| 3/2 4	6	12x2	8 △
{4,4/2} + {3,4}			6 □
Inscribed Octahedron			
5| 3/2 5	12	30x2	20 △
{5,5/2} + {3,5}			12 ⬠
Inscribed Icosahedron			

Table 12: More Quasi-Regular Polyhedra

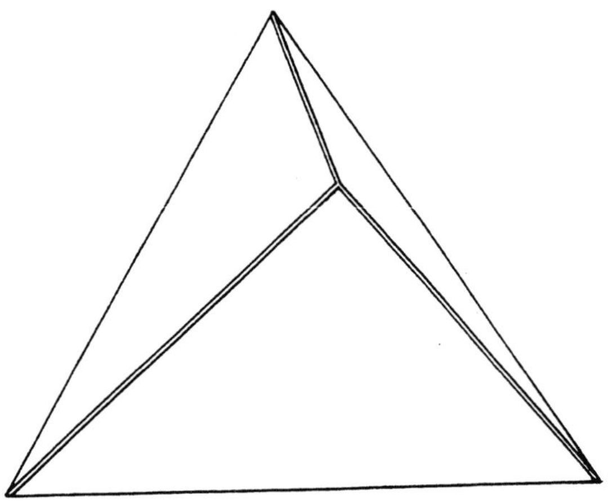

3| 3/2 3

{3,3/2} + {3,3}
Inscribed Tetrahedron

Fig.35: More Quasi-Regular Polyhedra

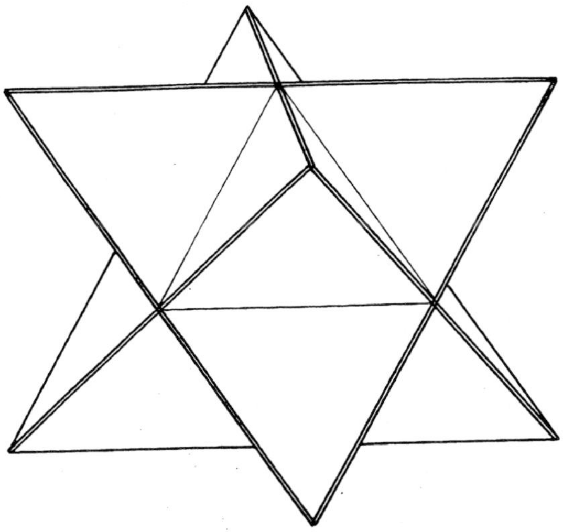

3| 4/2 3

(19) [1.4a]
Stella Octangula

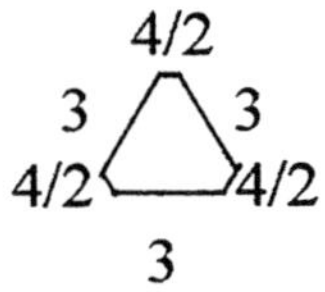

3| 5/2 3
39 (70)
Small
Ditrigonal Icosidodecahedron

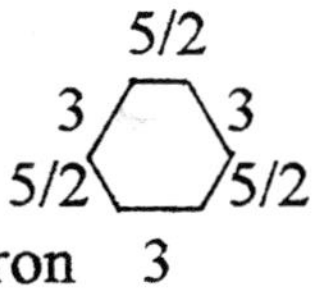

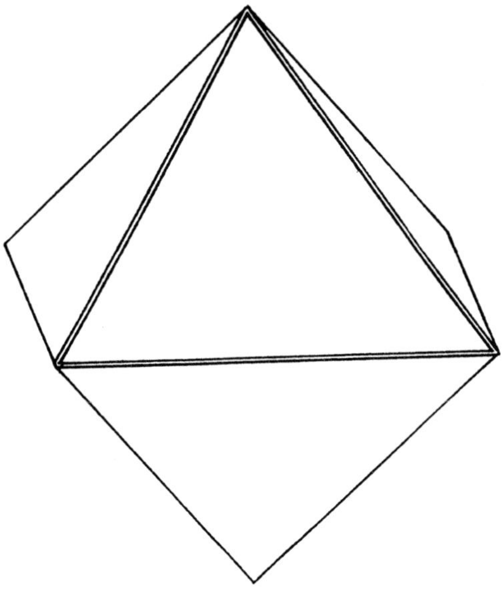

4| 3/2 4

{4,4/2} + {3,4}
Inscribed Octahedron

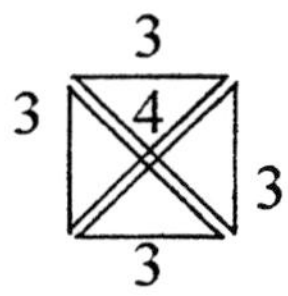

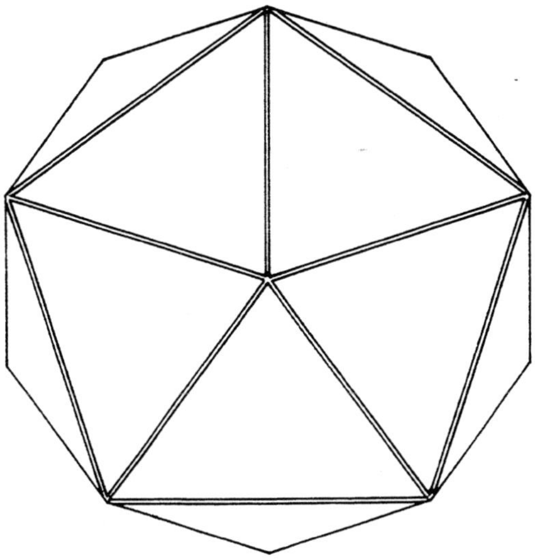

5| 3/2 5

{5,5/2} + {3,5}
Inscribed Icosahedron

MORE RHOMBIHEDRA

Moving on to the upper edge of our new Schwarz triangle we find x y/2 |x, a kind of rhombihedron with a trapezoid vertex figure. Here {x} and {y/2} are the 'parent' faces and two {2x}, the truncation of {x}, faces join them together.

These rhombihedra are shown in Figure 36 and described below in Table 13.

Again when y is 4 or 5, we get interesting new polyhedra with normal rhombihedral vertex figures, one the Truncated Stella Octangula; the other the somewhat similar looking Small Icosicosidodecahedron.

When y is 3, the rhombihedra are not without interest either. As before the {3/2} face puts a twist into the vertex figure, forming a quasirhombihedron type of polyhedron. These are in fact the previously seen p-type rhombihedra from our original parent families, here derived without resort to the missing out of faces.

All of these polyhedra again have double density and can be made to comply with our revised Euler's formula, V+F=E+4, by again counting the double density {y/2} faces twice.

Symbol	V	E	F
3 5/2 |3 40 (71) Small Icosicosidodecahedron	60	120	12 ☆ 20 △ 20 ⬡
3 4/2 |3 [1.4b] Truncated Stella Octangula	24	48	6 ✳ 8 △ 8 ⬡
3 3/2 |3 p{3 3} 37 (68) Octahemioctahedron	12	24	4 △ 4 △ 4 ⬡
4 3/2 |4 p{3 4} 38 (69) Small Cubicuboctahedron	24	48	8 △ 6 □ 6 ⯃
5 3/2 |5 p{3 5} 42 (72) Small Dodecicosidodecahedron	60	120	20 △ 12 ⬠ 12 ⯃

Table 13: More Rhombihedra

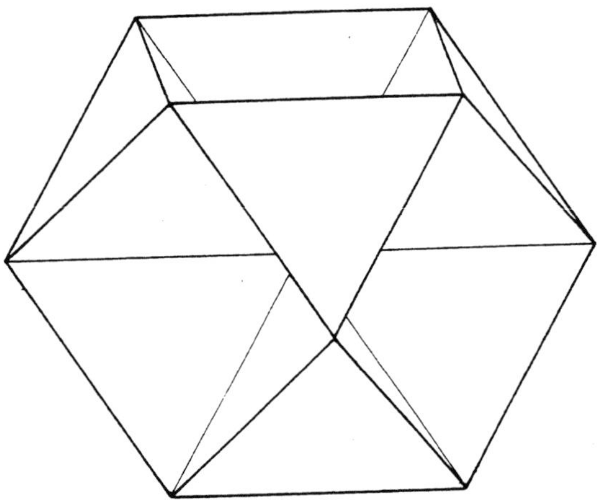

3 3/2 |3 p{3 3}
37 (68)
Octahemioctahedron

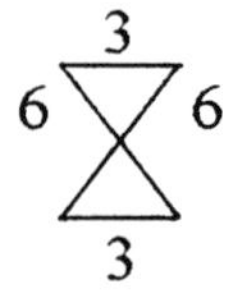

Fig.36: More Rhombihedra

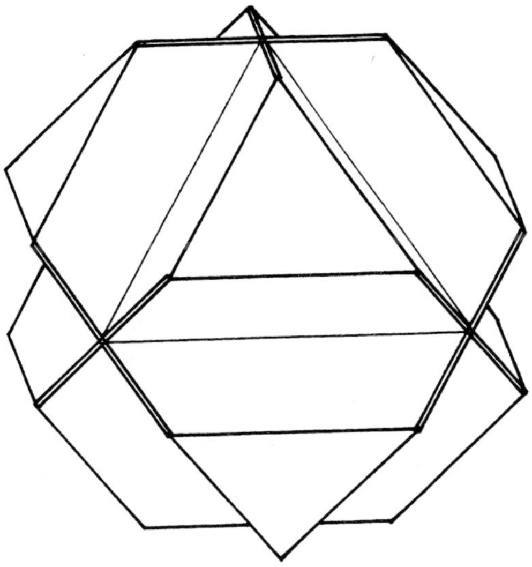

3 4/2 |3

[1.4b]
Truncated Stella Octangula

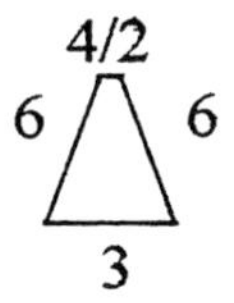

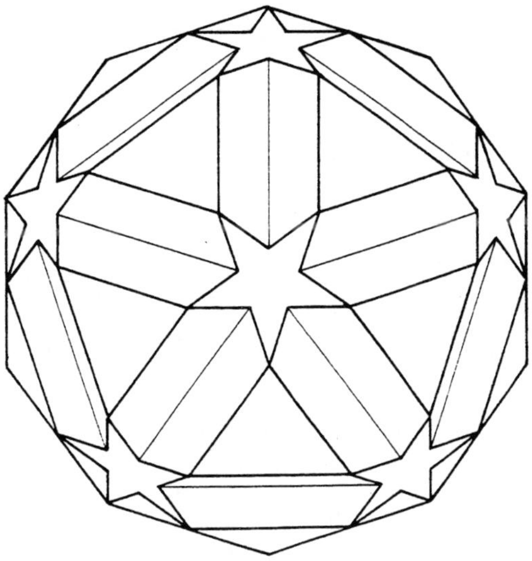

3 5/2 |3

40 (71)
Small Icosicosidodecahedron

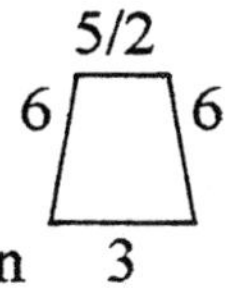

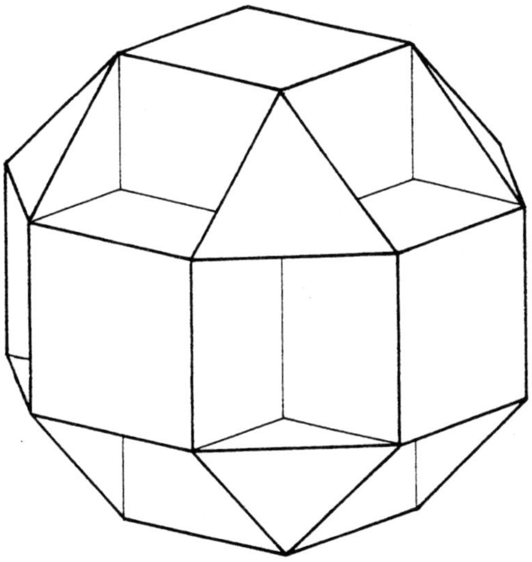

4 3/2 |4 $p\begin{Bmatrix}3\\4\end{Bmatrix}$

38 (69)
Small Cubicuboctahedron

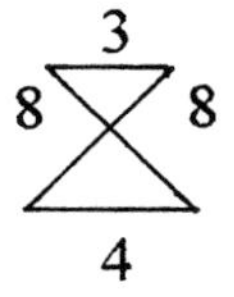

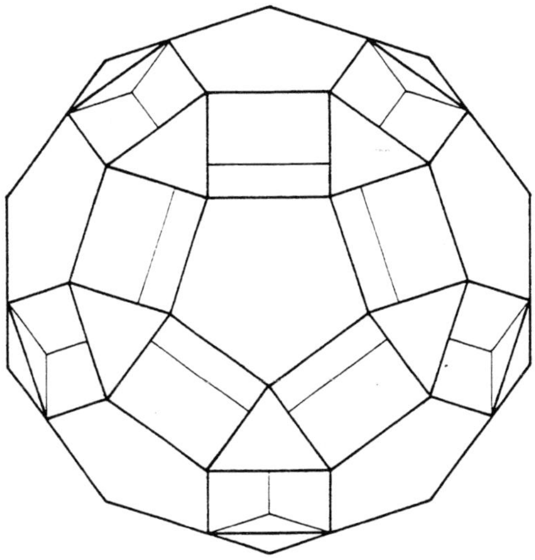

5 3/2 |5 $p\begin{Bmatrix}3\\5\end{Bmatrix}$

42 (72)
Small Dodecicosidodecahedron

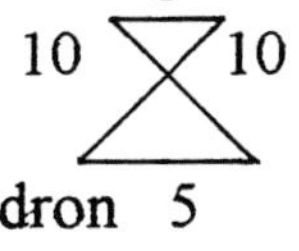

MORE EVEN FACED POLYHEDRA

It would seem that our efforts at exploring within our new Schwarz triangles are being rewarded, so we will now press on and take a look outside, firstly at the centres of our upper colunar triangles (x y/2' x').

Naturally enough these yield the symbol x (y/2)' x'|, a quasiquasitruncated even faced polyhedron with a triangular vertex figure. Note that since none of x, x' or y/2' are 2, we do not get any single square faces {4} resulting from the truncation.

However since all the faces in the vertex figure are truncations these polyhedra are still even faced as shown in Figure 37 and described below in Table 14.

Of all these, only the Quasiquasitruncated Inscribed Octahedron has a single triangular vertex figure. It was recognised as uniform by Coxeter et al. and given its official name of Cuboctatruncated Cuboctahedron.

The others all involve double polygons, the two with fivefold symmetry producing proper double triangle vertex figures (see p.56). These provide the shells for two additional rhombihedra, one which we have just seen as 3 5/2 |3, the other being 3 5 |5/3, which we shall not see until the next volume.

All of these double density polyhedra still comply with Euler's formula, V+F=E+4, this time by counting the t{y/2'} faces twice.

Symbol	V	E	F
3 5/3 3/2|	60x2	60x2	12
Quasiquasitruncated Small		60x1	20
Ditrigonal Icosidodecahedron			20
3 4/2 3/2|	12x4	24x3	6
Quasiquasitruncated			8
Stella Octangula			8
3 3 3/2|	12x2	18x2	4
Quasiquasitruncated			4
Inscribed Tetrahedron			4
4 3 4/3|	48	72	8
52 (79)			6
Cuboctatruncated Cuboctahedron			6
5 3 5/4|	60x2	60x2	20
Quasiquasitruncated		60x1	12
Inscribed Icosahedron			12

Table 14: More Even Faced Polyhedra

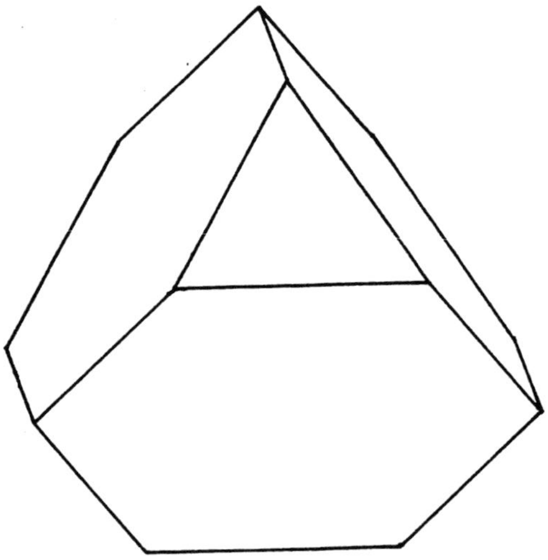

3 3 3/2|

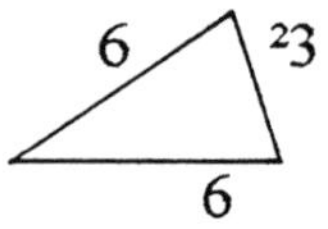

Quasiquasitruncated
Inscribed Tetrahedron

Fig.37: More Even Faced Polyhedra

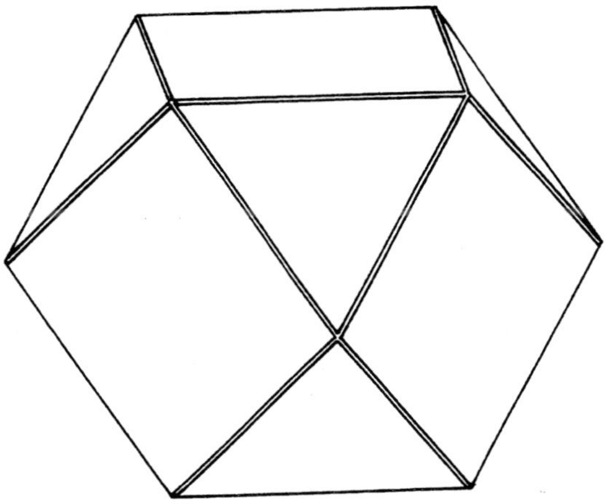

3 4/2 3/2|

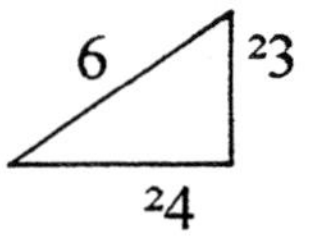

Quasiquasitruncated
Stella Octangula

3 5/3 3/2|

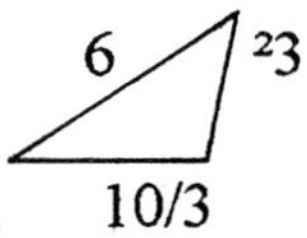

Quasiquasitruncated Small
Ditrigonal Icosidodecahedron

4 3 4/3|

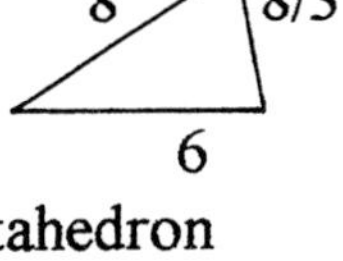

52 (79)
Cuboctatruncated Cuboctahedron

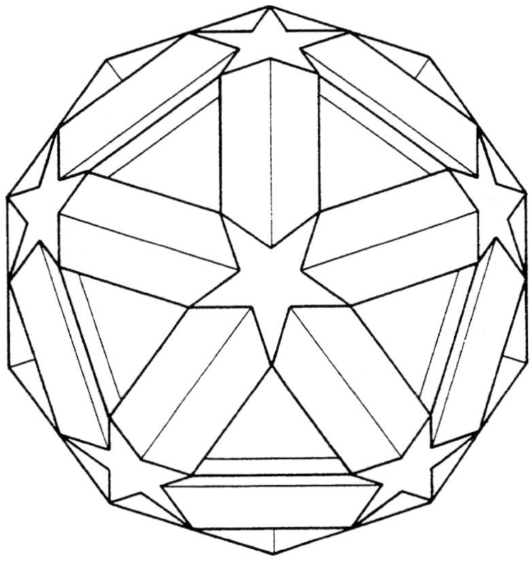

5 3 5/4|

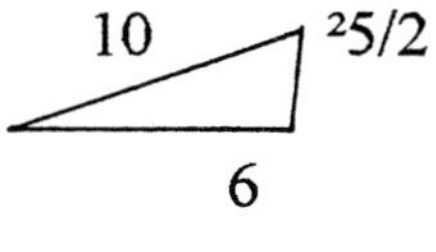

Quasiquasitruncated
Inscribed Icosahedron

MORE QUASIRHOMBIHEDRA

Regarding the previous set of even faced polyhedra as quasiquasitruncations we can continue the process in two different ways.

Firstly we will consider having started at the (x) vertex and continue the truncation via the even faced polyhedron to end up on the extended upper edge of the Schwarz triangle at x (y/2)' |x', a kind of quasirhombihedron.

This will have the usual trapezoid vertex figure, but with {x} and {y/2'} as the 'parent' faces along with two t{x'}, the quasitruncation of {x}, faces joining them together.

These are shown in Figure 38 and described below in Table 15.

Symbol		V	E	F
3 5/3 |3/2		12x5	30x4	12 ✯
{5/2,5} + 3{3,5/2}				20 △
Small Quasicosicosidodecahedron				20 ▲
3 4/2 |3/2		8x3	24x2	6 ✕
(19) [1.4a]				8 △
Double Stella Octangula				8 ▲
3 3 |3/2	p'{3 3}	6x2	12x2	4 △
17 (2)				4 △
Double Octahedron				4 ▲
4 3 |4/3	p'{3 4}	24	48	8 △
50 (77)				6 □
Great Cubicuboctahedron				6 ✴
5 3 |5/4	p'{3 5}	20x3	60x2	20 △
Small				12 ⬠
Quasidodecicosidodecahedron				12 ★

Table 15: More Quasirhombihedra

In parallel with the rhombihedra we saw on page 48, three of these include a triangle face, but now as the inverse of {3/2}, so that although technically quasirhombihedra they have rhombihedra type vertex figures without a twist.

These three are in fact the previously seen p'-type quasirhombihedra from our original parent families. The other two comprise a Double Stella Octangula and another inscribed version of what appears to be a {5/2,5}, similar to the Quasitruncated Dodecahedron.

By counting the {y/2'} faces twice, as before, these still comply with our revised Euler's formula for double density polyhedra, V+F=E+4.

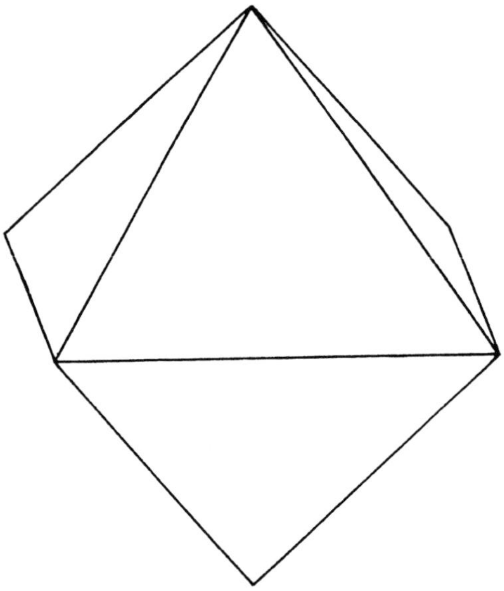

3 3 |3/2 p'{3 3}

17 (2)

Double Octahedron

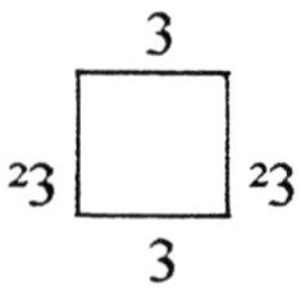

Fig.38: More Quasirhombihedra

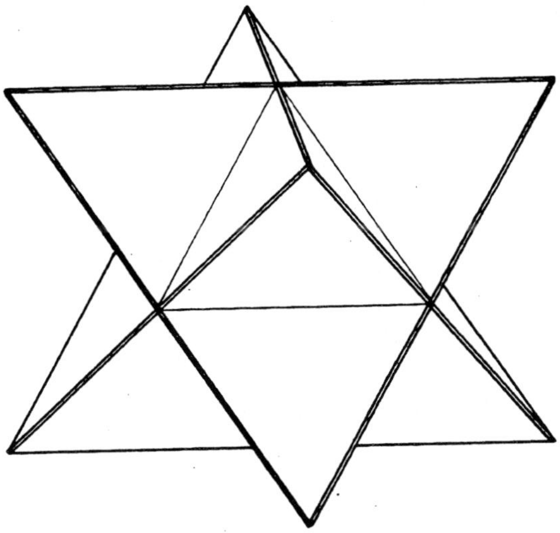

3 4/2 |3/2

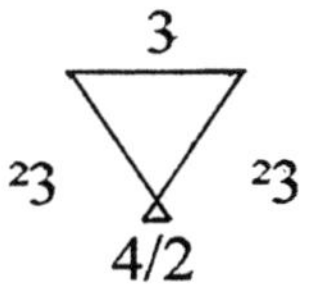

(19) [1.4a]
Double Stella Octangula

3 5/3 |3/2
{5/2,5} + 3{3,5/2}
Small
Quasicosicosidodecahedron

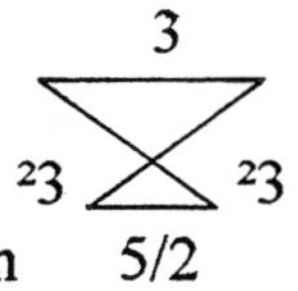

4 3 |4/3 $p'\begin{Bmatrix}3\\4\end{Bmatrix}$

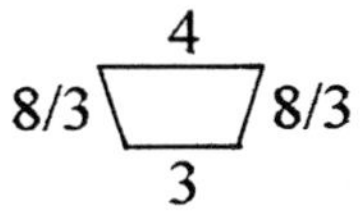

50 (77)
Great Cubicuboctahedron

5 3 |5/4 $p'\begin{Bmatrix}3\\5\end{Bmatrix}$

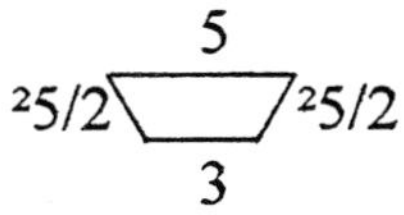

Small Quasidodecicosidodecahedron

SOME HEMIHEDRA

Taking the second route from the (y/2) vertex, via the quasiquasitruncated even faced polyhedron, we arrive at a rhombihedron on the extended lower edge of our Schwarz triangle. This is x x' |(y/2)', the polyhedral equivalent of the coincident tilings, and has the inverse pair {x} and {x'} as 'parent' faces joined together by a pair of t{y/2'} faces.

Twisting the vertex figure to accomodate the inverse polygons, we arrive at a quasirhombihedron type vertex figure with two pairs of equal faces, two opposite and two diagonal. These are the hemihedra, p-type versions of the quasi-regular polyhedra, as shown in Figure 39 and described in Table 16 below.

The diagonal faces pass through the centre of each polyhedron cutting it exactly in half. In order for these to appear facing out both ways, it is necessary to double up the vertex figures as shown, twice where we have {²4} faces. This doubling is not required for the Octahemioctahedron, as the hexagons there are tetrahedral rather than co-planar.

The two hemihedra shown with five-fold symmetry both have {5/2} as their missing face. They thus appear to be derived from quasi-regular polyhedra from our suspected {3,5/2} and {5,5/2} families.

As might by now be expected the hemihedra have double density and can be made to comply with Euler's formula, V+F=E+4, by counting the t{y/2'} faces twice.

Symbol		V	E	F
3 3/2 |5/3	$h\begin{Bmatrix}3'\\5/2\end{Bmatrix}$	30x2	60x2	12 ✶
85 (106)				20 △
Great Icosihemidodecahedron				20 △
3 3/2 |4/2	$h\begin{Bmatrix}3'\\3\end{Bmatrix}$	6x4	12x4	6 ⊡
36 (67)				8 △
Double Tetrahemihexahedron				8 △
3 3/2 |3	$h\begin{Bmatrix}3'\\4\end{Bmatrix}$	12	24	4 ⬡
37 (68)				4 △
Octahemioctahedron				4 △
4 4/3 |3	$h\begin{Bmatrix}3\\4'\end{Bmatrix}$	12x2	24x2	8 ⬡
51 (78)				6 □
Cubohemioctahedron				6 □
5 5/4 |3	$h\begin{Bmatrix}5'\\5/2\end{Bmatrix}$	30x2	60x2	20 ⬡
81 (102)				12 ⬠
Great Dodecahemicosahedron				12 ⬠

Table 16: Some Hemihedra

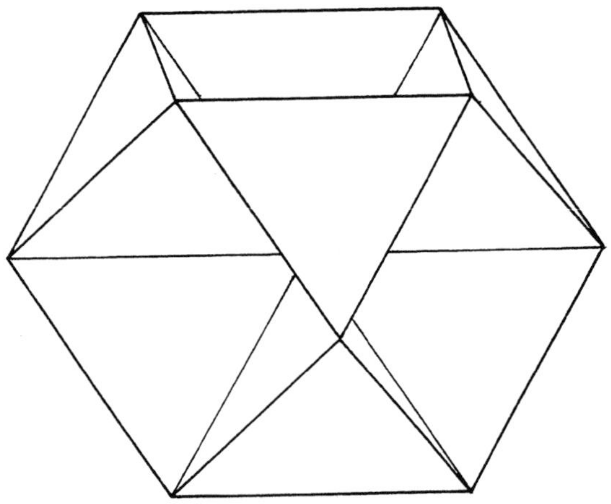

3 3/2 |3 $h\begin{Bmatrix}3'\\4\end{Bmatrix}$

37 (68)

Octahemioctahedron

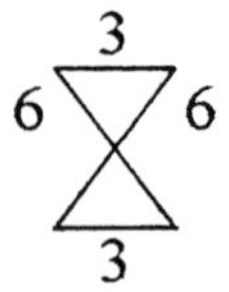

Fig.39: Some Hemihedra

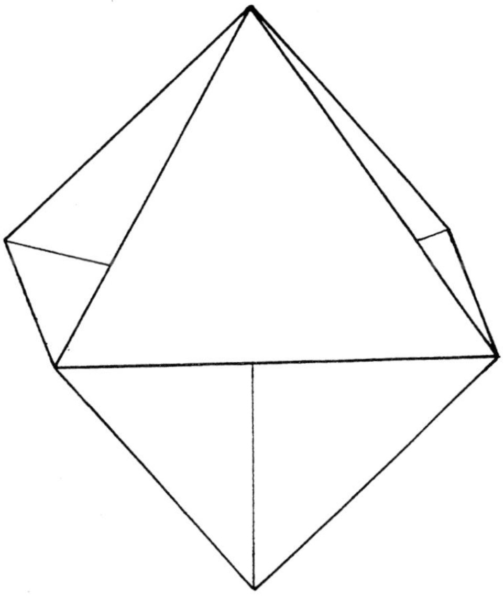

3 3/2 |4/2 $h\begin{Bmatrix}3'\\3\end{Bmatrix}$

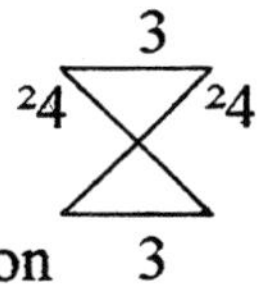

36 (67)

Double Tetrahemihexahedron

3 3/2 |5/3 $h\begin{Bmatrix}3'\\5/2\end{Bmatrix}$

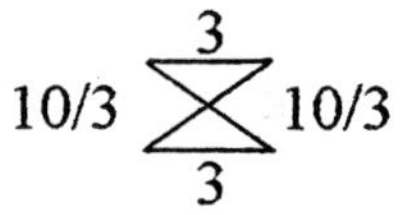

85 (106)

Great Icosihemidodecahedron

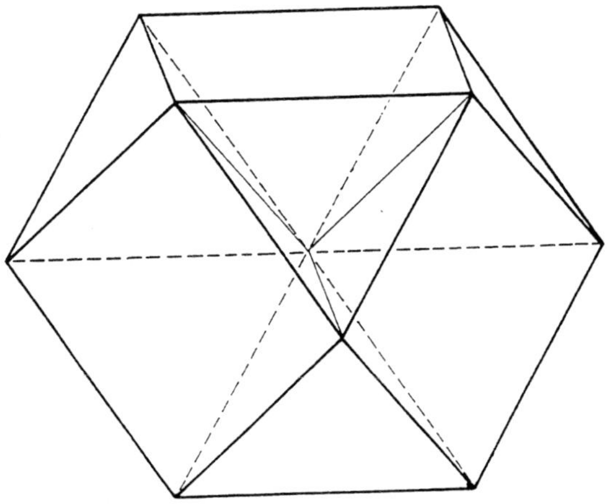

4 4/3 |3 $h\begin{Bmatrix}3\\4'\end{Bmatrix}$

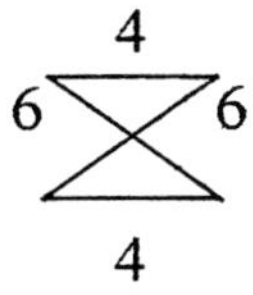

51 (78)

Cubohemioctahedron

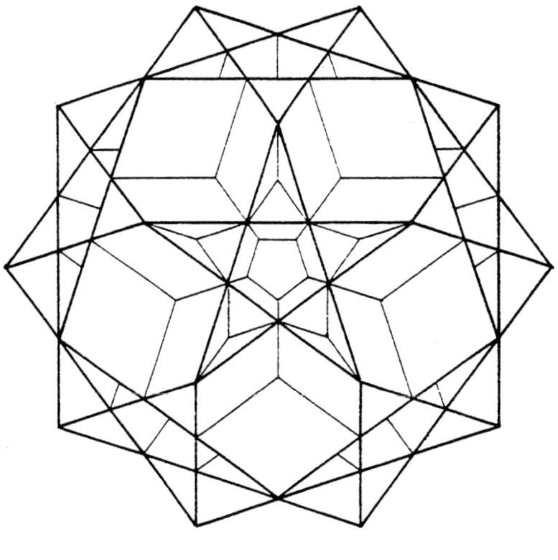

5 5/4 |3 $h\begin{Bmatrix}5'\\5/2\end{Bmatrix}$

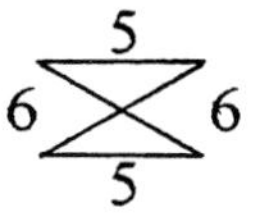

81 (102)

Great Dodecahemicosahedron

MORE DOUBLE POLYGON VERTICES

Some of the sub-family polyhedra we have been examining have produced new vertex figures with double polygons. These can be treated in exactly the same way as those we saw earlier, the double polygons becoming single ones present twice, so that the triangular vertices of even faced polyhedra double up to fit within trapezoidal shells.

Three of these are shown on the right in Figure 40. The first, with two double polygons, produces a double inscribed cuboctahedron vertex, whilst the second fits within the shell of the recently seen 3 5/2 |3. The quasirhombi form that fits within this last shell is 3 5/3 |5, belonging to a possible (3 5/3 5) super-family, as does the 3 5 |5/3 shell that surrounds our third example.

The remaining four vertex figures shown below are quasirhombihedra with paired double polygon vertices, which can be built up into multiple vertices in a similar way.

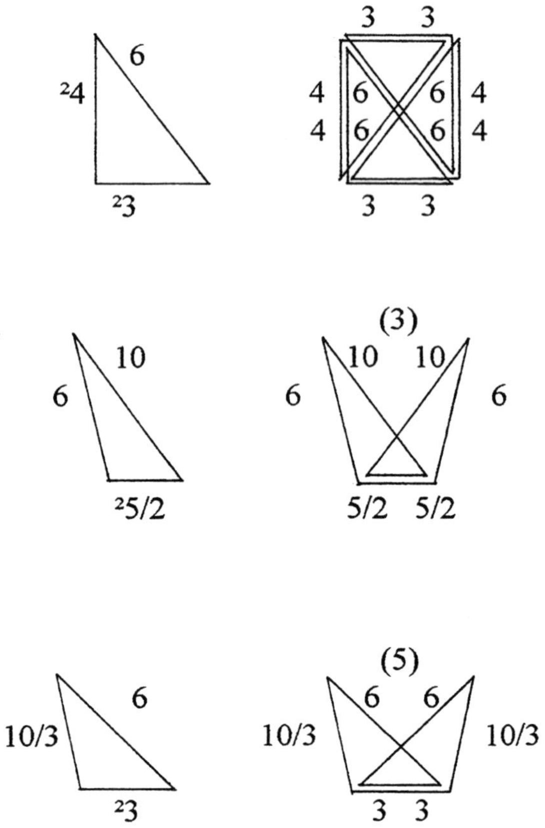

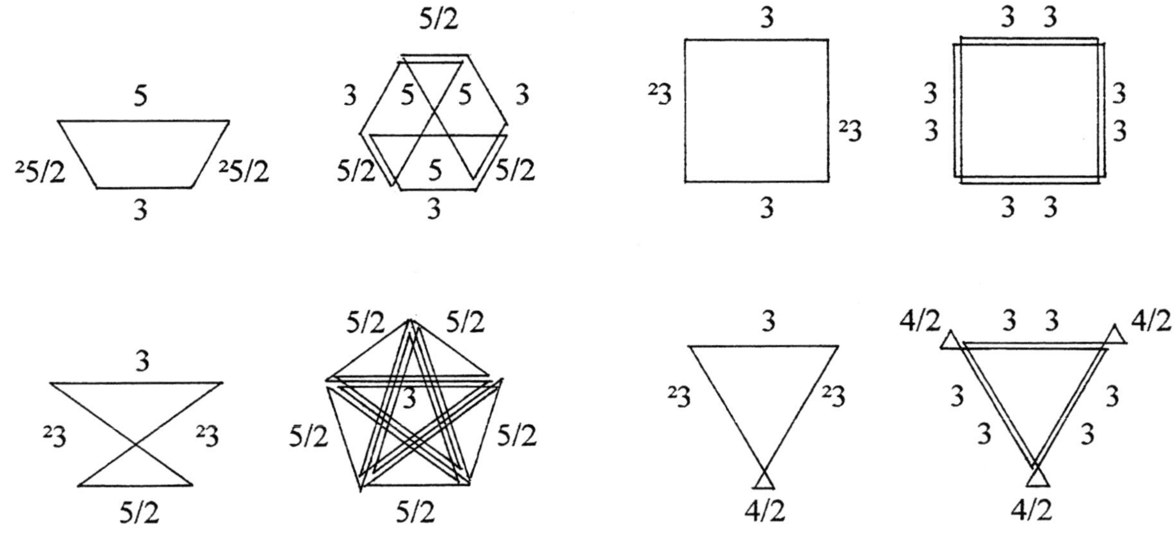

Fig.40: More Double Polygon Vertices

MORE P- & Q-TYPE RHOMBIHEDRA

In the past few pages we have also seen a number of new rhombihedra with trapezoid vertex figures. These also deserve examination for p- or q-type derivations formed by taking the six faces present and missing out different pairs.

Obviously those that are already the p- and p'-type rhombihedra from our parent symmetry families will only yield back the parent rhombihedra and the q-types we have already seen.

This leaves us with the uppermost pairs of rhombihedra and quasirhombihedra on pages 49 and 53, which should yield a total of eight more polyhedra. Five of these are found to be further inscribed versions of what we have already seen, whilst the other three deserve mention and are illustrated below in Figure 41.

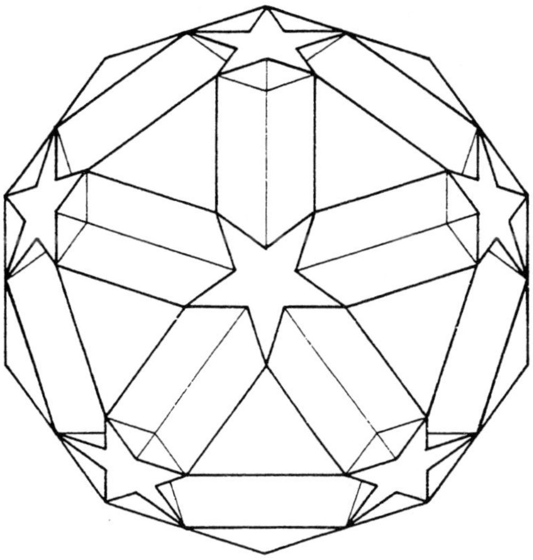

3 5/3 |5
55 (82)
Small
Ditrigonal Dodecicosidodecahedron

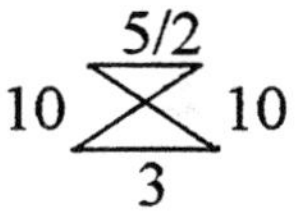

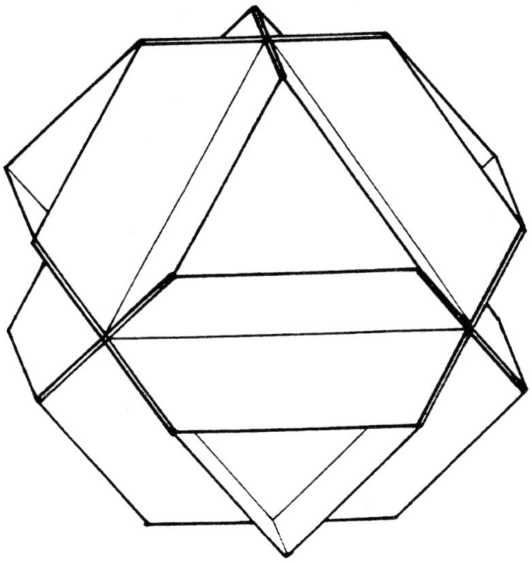

3 3 3/2|
4/4|

Octoctahedron

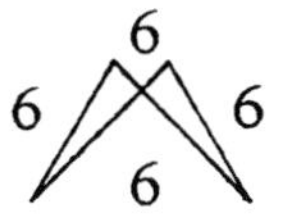

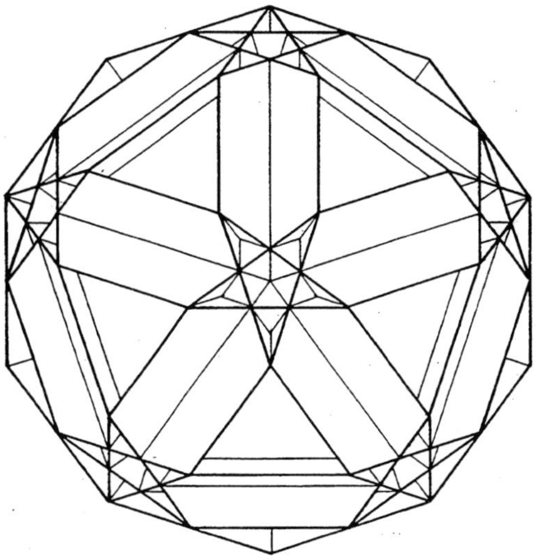

3 5 3/2|
5/4|
64 (90)
Small Dodecicosahedron

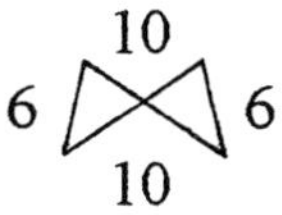

Fig.41: More P- & Q-type Rhombihedra

MORE SNUB POLYHEDRA

The two uppermost rhombihedra on page 49 had paired hexagon faces, which can be taken as the basis for further snub polyhedra in our new sub-families. By removing the two hexagons, applying a twist to the vertices and replacing them with a triangle and three triangles respectively, we can obtain the two hexisnub polyhedra with the general symbol |3 y/2 3.

These appear uppermost in Figure 42 opposite and both have double triangular star faces where two triangles share the same plane. The Snub Stella Octangula consists of a pair of interlocked icosahedra linked together by {4/2} faces whilst the Small Snub Icosicosidodecahedron has its star faces connected by {5/2} faces.

Generalising further we can construct snub versions of the other rhombihedra on page 49 by alternating triangles and the polygons represented in the sub-family symbol around a hexagonal vertex figure. We thus obtain |x y/2 x with a sequence around each vertex of (3, x, 3, y/2, 3, x).

For the lower three shown in Figure 42 the predicted {3/2} face appears as a {3} plus a twist, as it did for the rhombihedra. These thus appear to be versions of the quasi-regular polyhedra with double vertices and threefold edges. All five of these snub polyhedra are described in Table 17 below. Again all have double density and comply with Euler's formula, V+F=E+4, when the {y/2} faces are counted twice.

Symbol	V	E	F	
|3 5/2 3	60	180	12	✯
41 (110)			40	✡
Small Snub Icosicosidodecahedron			60	△
|3 4/2 3	24	72	6	✳
[1.4c]			16	✡
Snub Stella Octangula			24	△
|3 3/2 3	6x2	12x3	4	△
17 (2)			8	△
Octahedron			12	△
|4 3/2 4	12x2	24x3	8	△
19 (11)			12	□
Cuboctahedron			24	△
|5 3/2 5	30x2	60x3	20	△
28 (12)			24	⬠
Icosidodecahedron			60	△

Table 17: More Snub Polyhedra

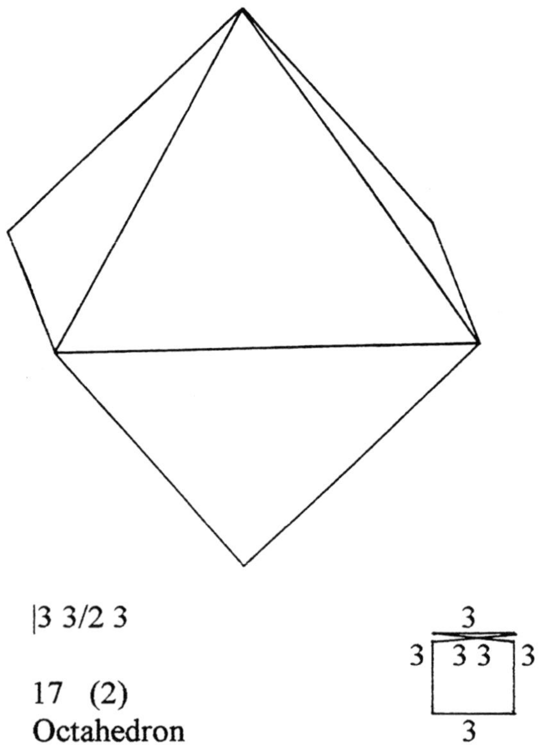

Fig.42: More Snub Polyhedra

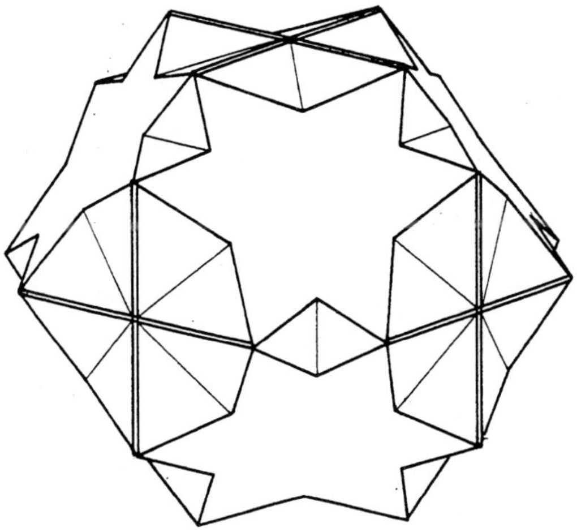

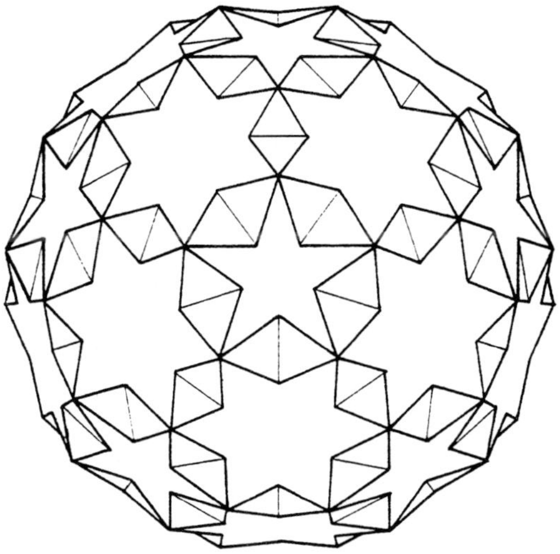

|3 4/2 3

[1.4c]
Snub Stella Octangula

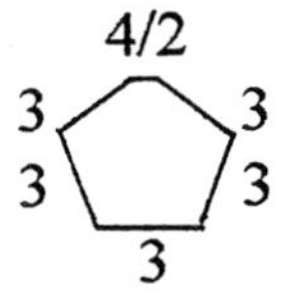

|3 5/2 3
41 (110)
Small
Snub Icosicosidodecahedron

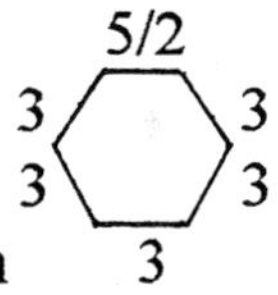

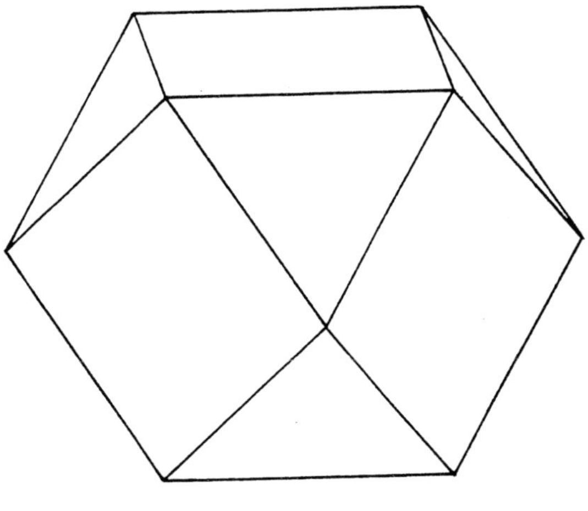

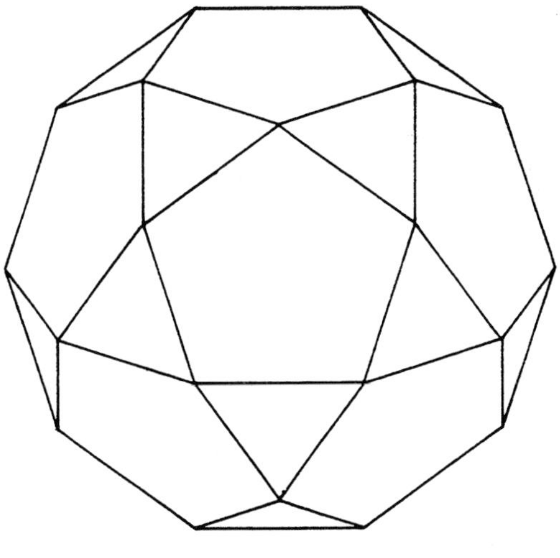

|4 3/2 4

19 (11)
Cuboctahedron

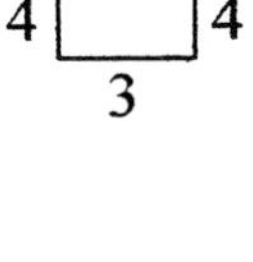

|5 3/2 5

28 (12)
Icosidodecahedron

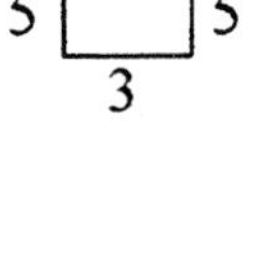

MORE QUASISNUB POLYHEDRA

Our original snub polyhedra were derived from the Archimedean rhombihedra by omitting square faces, then twisting and inserting two triangles. If we now look at them in the light of the last section we can see that the generalised rule for construction just used also works.

By including the digon {2} indicated by the symbol |x 2 y we can still produce a nominally hexagonal vertex figure, which because of the theoretical zero size of the digon matches the pentagonal vertex figure originally found for the Archimedeans.

The quasisnub polyhedra originally found can also be seen to contain a redundant digon. These new forms of vertex figure for both the snubs and quasisnubs from our original parent families are shown adjoining in Figure 43.

Applying this new technique of alternating triangles to the sub-family quasirhombihedra or hemihedra, we can obtain further quasisnub polyhedra of the general form |x (y/2)' x'. One of these, |3 3 3/2, we have already seen as |3 3/2 3, whilst another, |5 3 5/4, can only be constructed with a double twist making it equivalent to |5/4 3/2 5/4 (see next section). The remaining three are all polyhedra we have seen before in other guises and are shown in Figure 44 and described in Table 18 facing.

It should be noted that the Quasisnub Stella Octangula shown here is the correct one! That described in the author's own '*Additions to the Uniform Polyhedra*' as |3/2 4/2 3/2 is more correctly a Retrosnub Stella Octangula, indeed the |3/2 2 3/2 described by Coxeter et al. as a Quasisnub Tetrahedron should also be a retrosnub (qv).

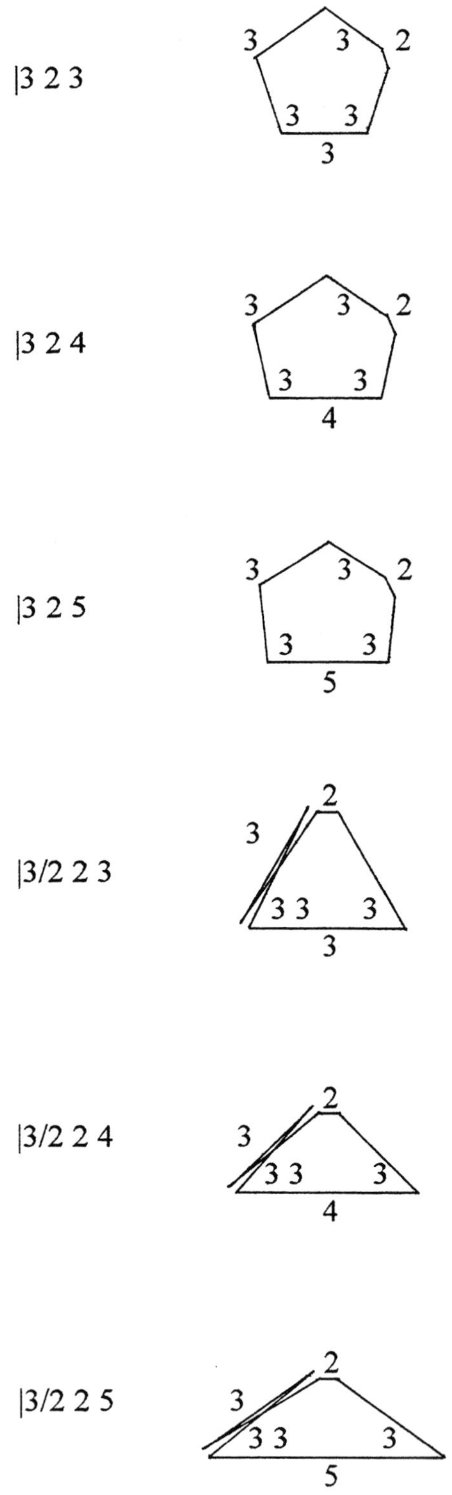

Fig.43: Snub & Quasisnub Vertex Figures

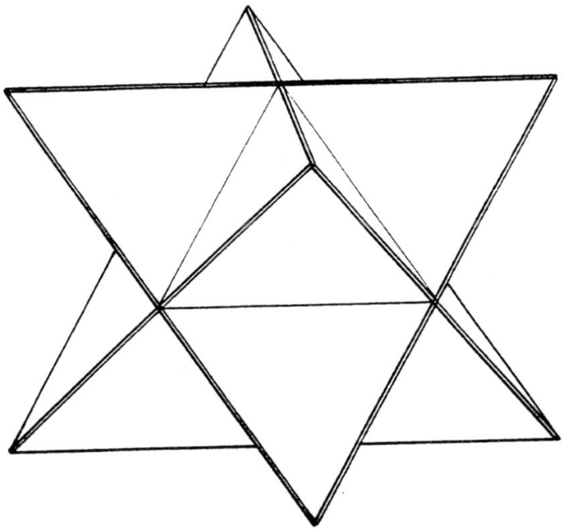

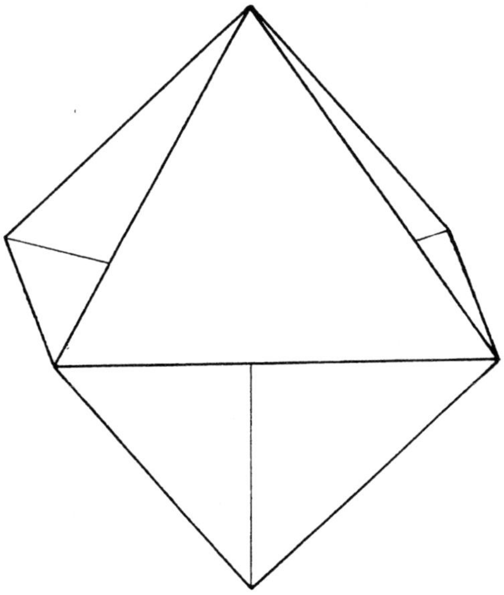

|3 4/2 3/2

(19) [1.4a]

Quasisnub Stella Octangula

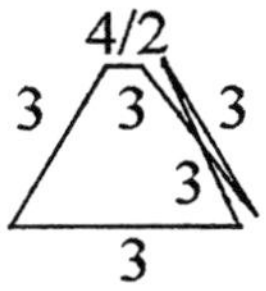

|3 5/3 3/2

{5/2,5} + 5{3,5/2}

Quasisnub

Icosicosidodecahedron

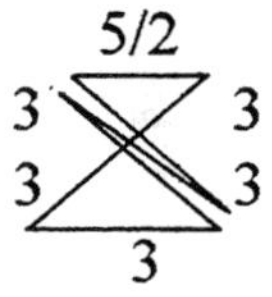

|4 3 4/3

36 (67)

Double Tetrahemihexahedron

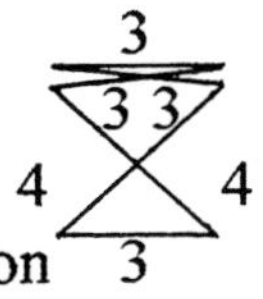

Fig.44: More Quasisnub Polyhedra

Symbol	V	E	F
|3 5/3 3/2	12x5	30x6	12 ☆
{5/2,5} + 5{3,5/2}			40 △
Quasisnub Icosicosidodecahedron			60 △
|3 4/2 3/2	8x3	24x3	6 ✕
(19) [1.4a]			16 △
Quasisnub Stella Octangula			24 △
|3 3 3/2	6x2	12x3	4 △
17 (2)			8 △
Octahedron			12 △
|4 3 4/3	6x4	12x6	8 △
36 (67)			12 □
Double Tetrahemihexahedron			24 △

Table 18: More Quasisnub Polyhedra

RETROSNUB POLYHEDRA

We saw back on page 30 how theoretical double inverse faced quasi-regular polyhedra could be truncated to produce a quasiquasitruncated even faced polyhedron. Continuing the truncation led back to the standard rhombihedron as the paired inverse faces twisted the trapezoid vertex figure twice, thereby cancelling each other out. Starting with similar double inverse forms of the sub-family quasirhombihedra, we can generate a hexagonal snub vertex with inverses that will not cancel out. These retrosnub polyhedra, |x' y/2 x', have vertex figures with a double twist as shown in Figure 45 and described below in Table 19.

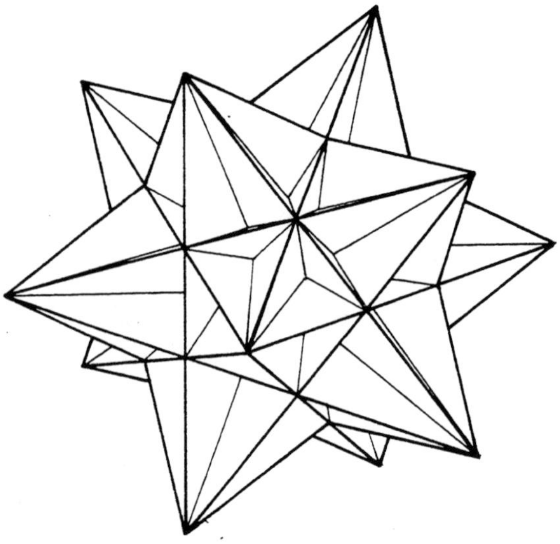

Symbol	V	E	F
|3/2 5/2 3/2	60	180	12 ✫
91 (118)			40 △
Sm I RSnub Icosicosidodecahedron			60 △
|3/2 4/2 3/2	24	72	6 ✕
[1.4d]			16 △
Retrosnub Stella Octangula			24 △
|3/2 2 3/2 {3,5/2}	12	30x2	30 =
69 (41)			8 △
Retrosnub Tetrahedron			12 △
|3/2 3/2 3/2	4x3	6x6	4 △
2{3,3/2} + 4{3,3}			8 △
Inscribed Tetrahedron			12 △
|4/3 3/2 4/3	6x4	12x6	8 △
2{4,4/2} + 4{3,4}			12 □
Inscribed Octahedron			24 △
|5/4 3/2 5/4	12x5	30x6	20 △
2{5,5/2} + 4{3,5}			24 ⬠
Inscribed Icosahedron			60 △

Table 19: Retrosnub Polyhedra

|3/2 2 3/2
{3,5/2}
69 (41)
Retrosnub Tetrahedron

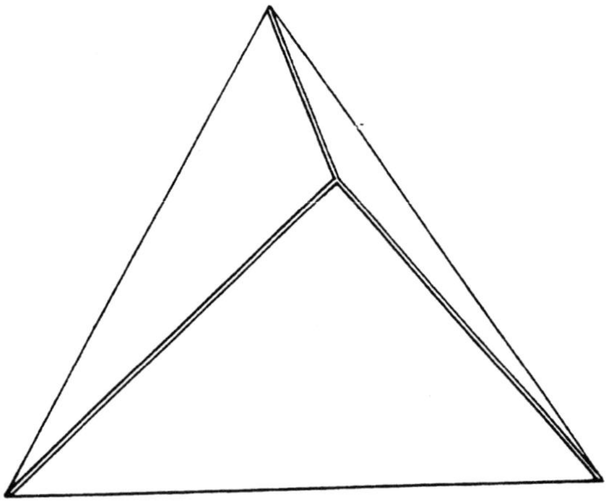

|3/2 3/2 3/2

2{3,3/2} + 4{3,3}
Inscribed Tetrahedron

Fig.45: Retrosnub Polyhedra

|3/2 4/2 3/2

[1.4d]
Retrosnub Stella Octangula

|3/2 5/2 3/2
91 (118)
Small Inverted Retrosnub
Icosicosidodecahedron

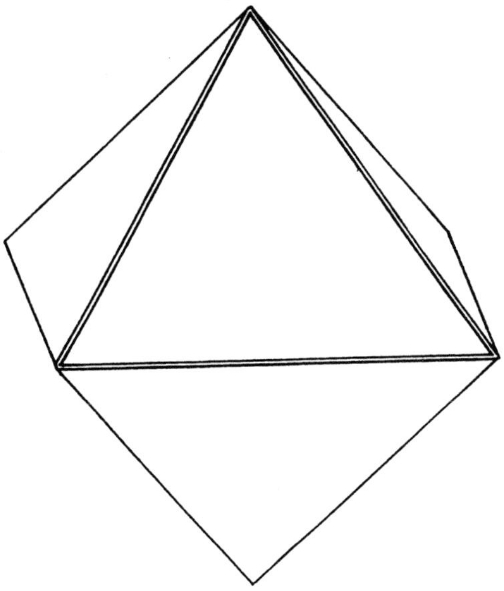

|4/3 3/2 4/3

2{4,4/2} + 4{3,4}
Inscribed Octahedron

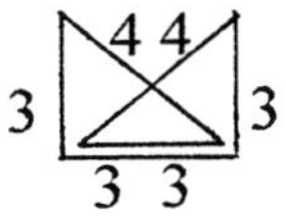

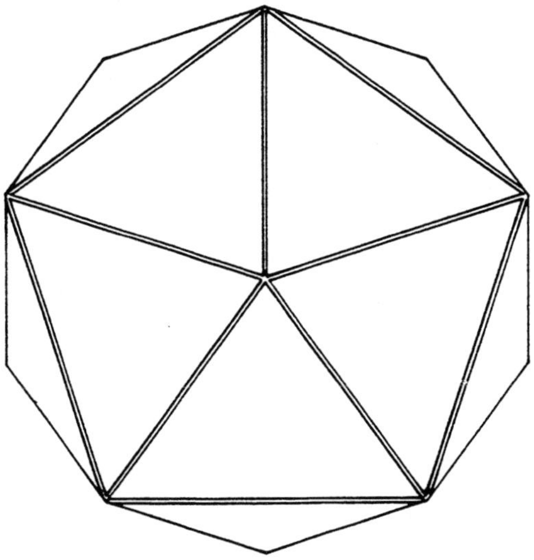

|5/4 3/2 5/4

2{5,5/2} + 4{3,5}
Inscribed Icosahedron

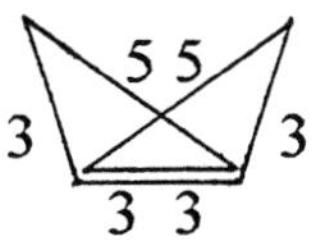

FIVE SUB-FAMILIES

We have now explored thoroughly our five sub-families of the general form (x y/2 x). Their Schwarz triangles and associated colunar triangles are shown adjoining in Figures 46, 47, 48, 49 and 50.

In general there are two types of 'quasi-regular' polyhedron in each sub-family: a double polyhedron (y/2| x x) and the quasi-regular polyhedron (x| y/2 x) as shown in Figure 35. There are three 'even faced' polyhedra: a double truncated parent polyhedron (x y/2 x|), its quasitruncated equivalent (x' y/2 x'|) and the new quasi-quasitruncated even faced polyhedron (x y/2' x'|) as shown in Figure 37.

In addition there are four 'rhombihedra' in each sub-family: a double quasi-regular polyhedron (x x |y/2), the hemihedron (x x' |y/2') as shown in Figure 39, the rhombihedron (x y/2 |x) as shown in Figure 36 and the quasirhombihedron (x y/2' |x') as shown in Figure 38.

Along the way we have had tantalising glimpses of further treasures in store belonging to such hinted at families as {3,5/2}, {5,5/2} and {4,4/2}. These are the subject of the next volume in this series specifically on the Star and Cross Polyhedra.

When we get there we will hopefully appreciate the description 'Simpler' in the title of the present volume, which will now be rounded off by a brief appendix dealing with the definitely simpler prisms and other related dihedra.

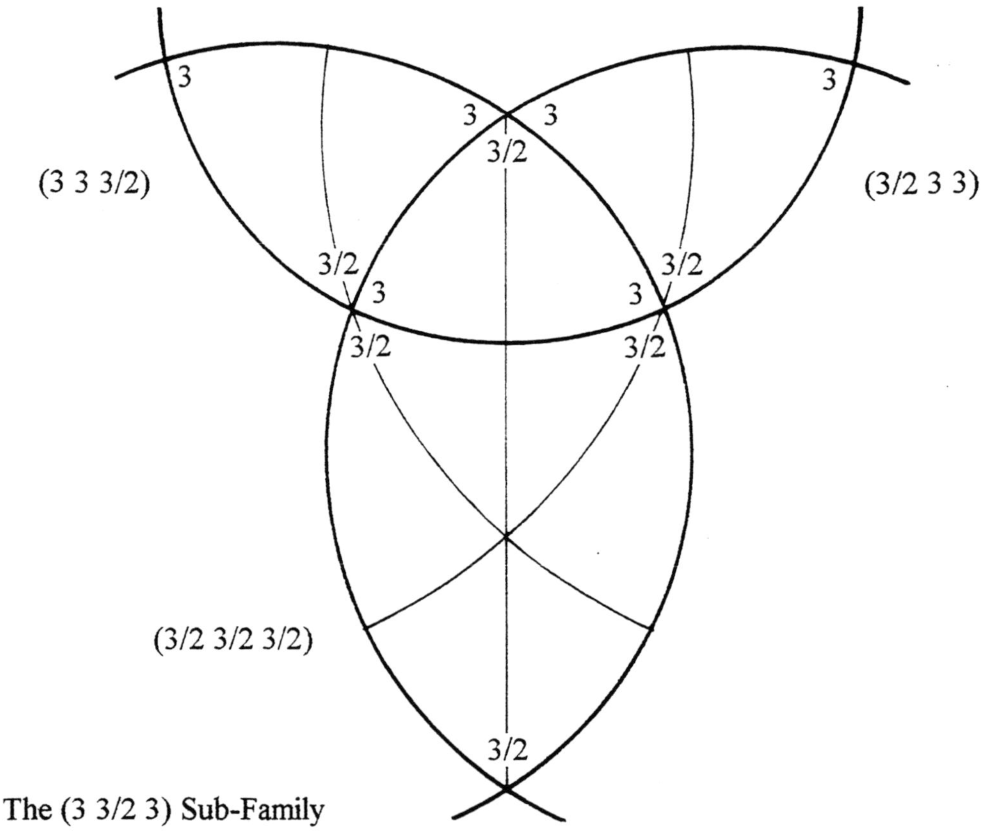

Fig.46: The (3 3/2 3) Sub-Family

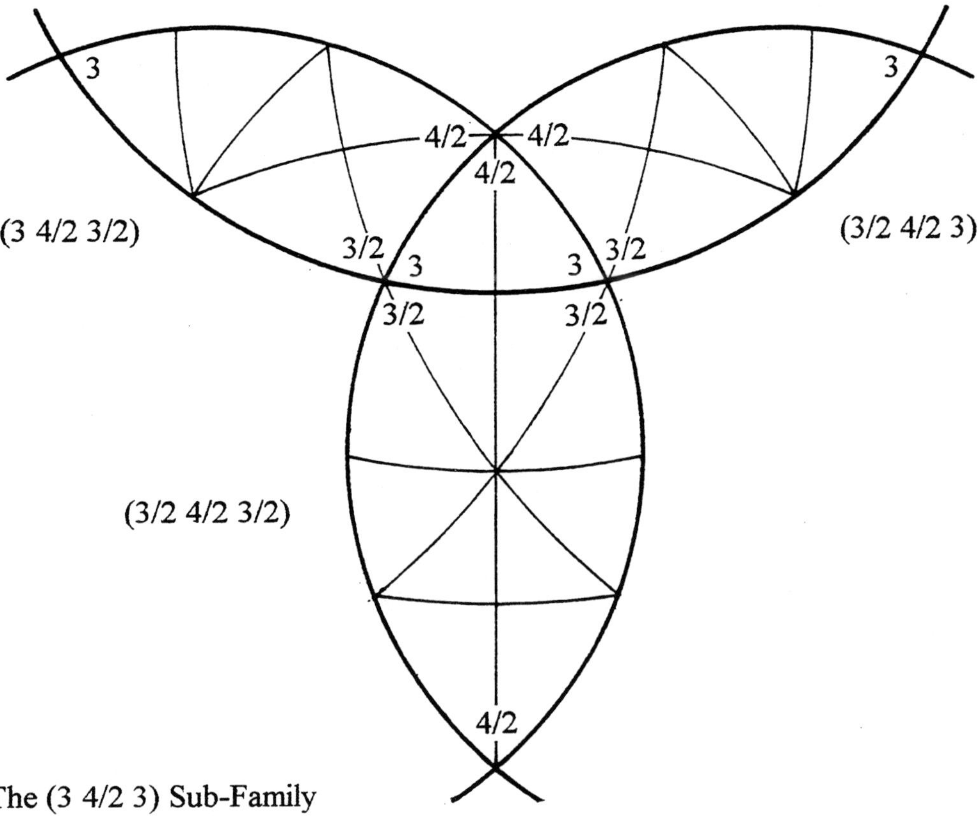

Fig.47: The (3 4/2 3) Sub-Family

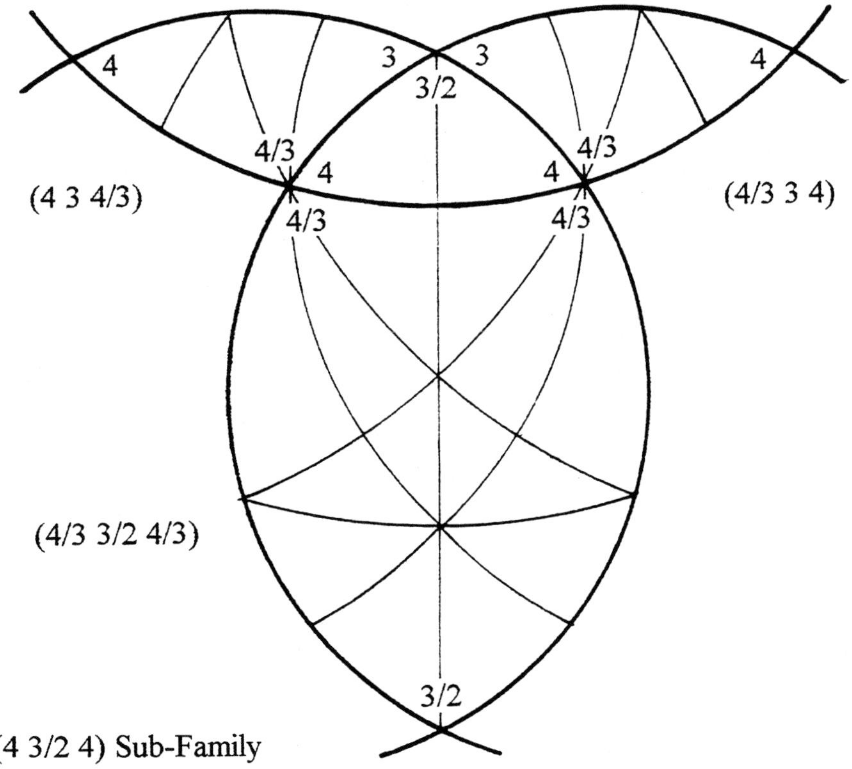

Fig.48: The (4 3/2 4) Sub-Family

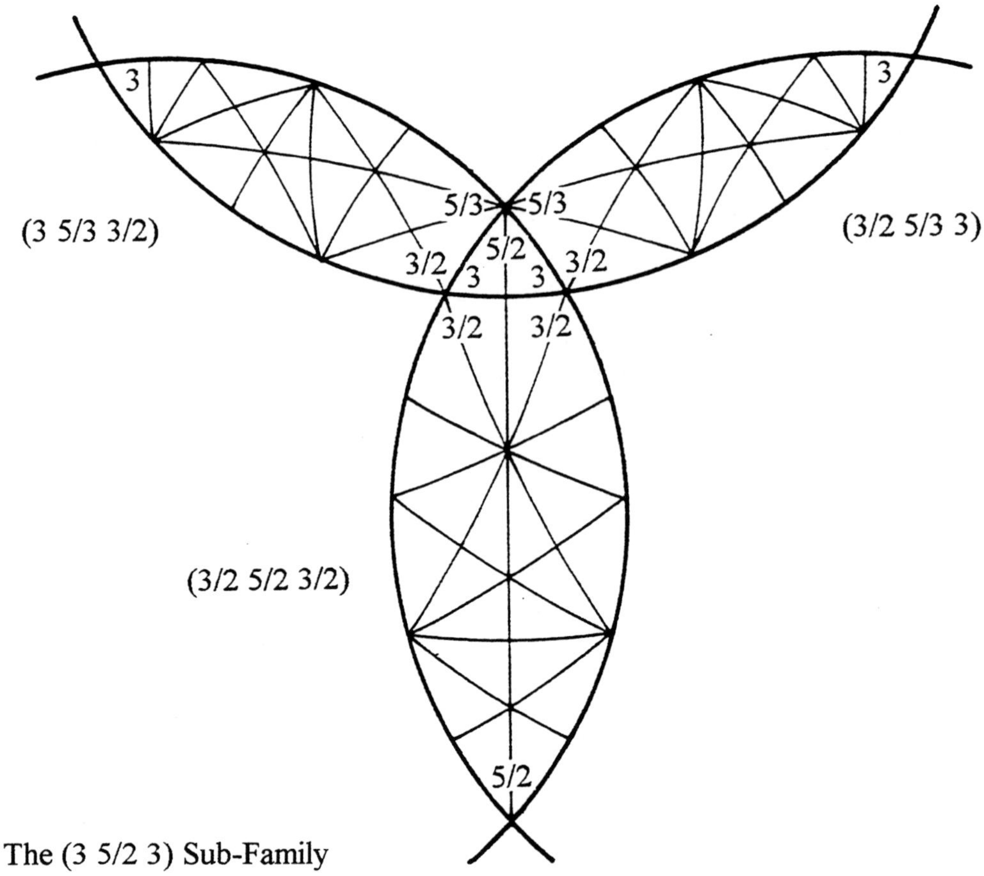

Fig.49: The (3 5/2 3) Sub-Family

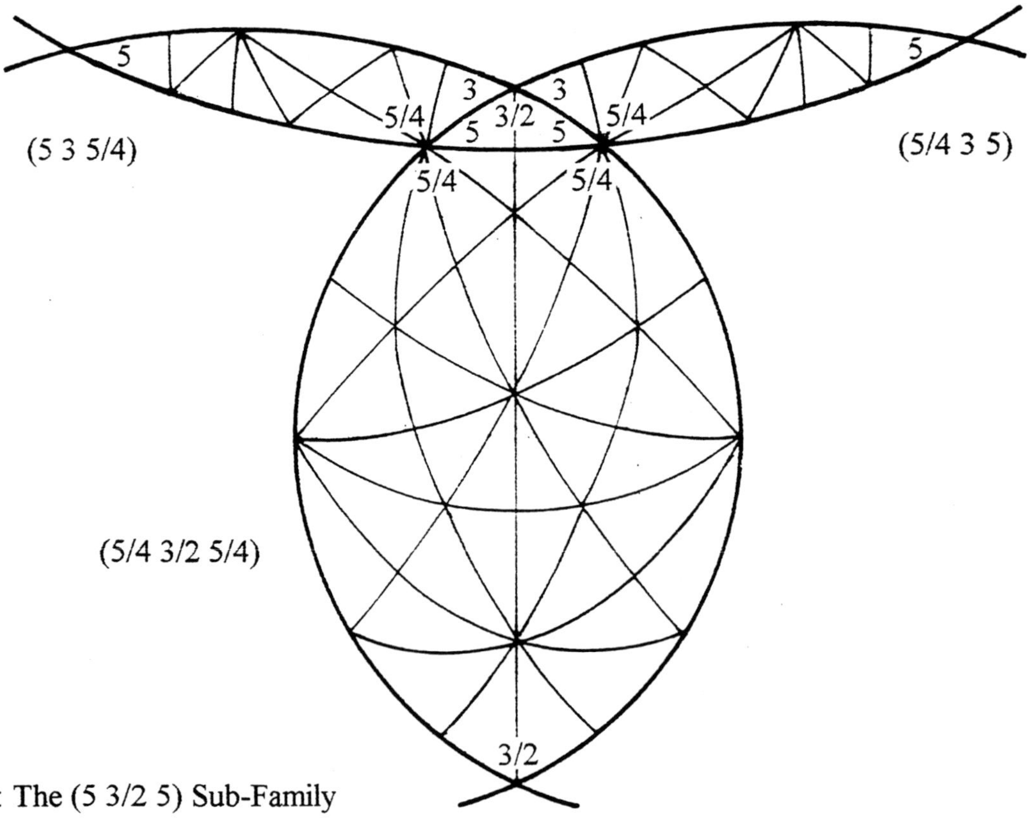

Fig.50: The (5 3/2 5) Sub-Family

REGULAR DIHEDRA

At the start of this volume we postponed considering the results of reducing the number of polygons per vertex to less than three. The time has now come to examine this proposition and in general we obtain {n,2}, a 'polyhedron' consisting of two {n} polygons back to back: 2| 2 n.

These dihedra have length and area but no volume, a property that excludes them from full membership of the uniform polyhedra. They do, however, have n vertices, n edges and 2 faces and thus still obey Euler's formula: V+F=E+2.

They form an infinite series of dihedra, the first five of which are shown in Figure 51 and described in Table 20 below.

Symbol		V	E	F
2| 2 2	{2,2}	2	2	2 ‖
Double Digon				
2| 2 3	{3,2}	3	3	2 △
Double Triangle				
2| 2 4	{4,2}	4	4	2 □
Double Square				
2| 2 5	{5,2}	5	5	2 ⬠
Double Pentagon				
2| 2 6	{6,2}	6	6	2 ⬡
Double Hexagon				

Table 20: Regular Dihedra

APPENDIX: DIHEDRA

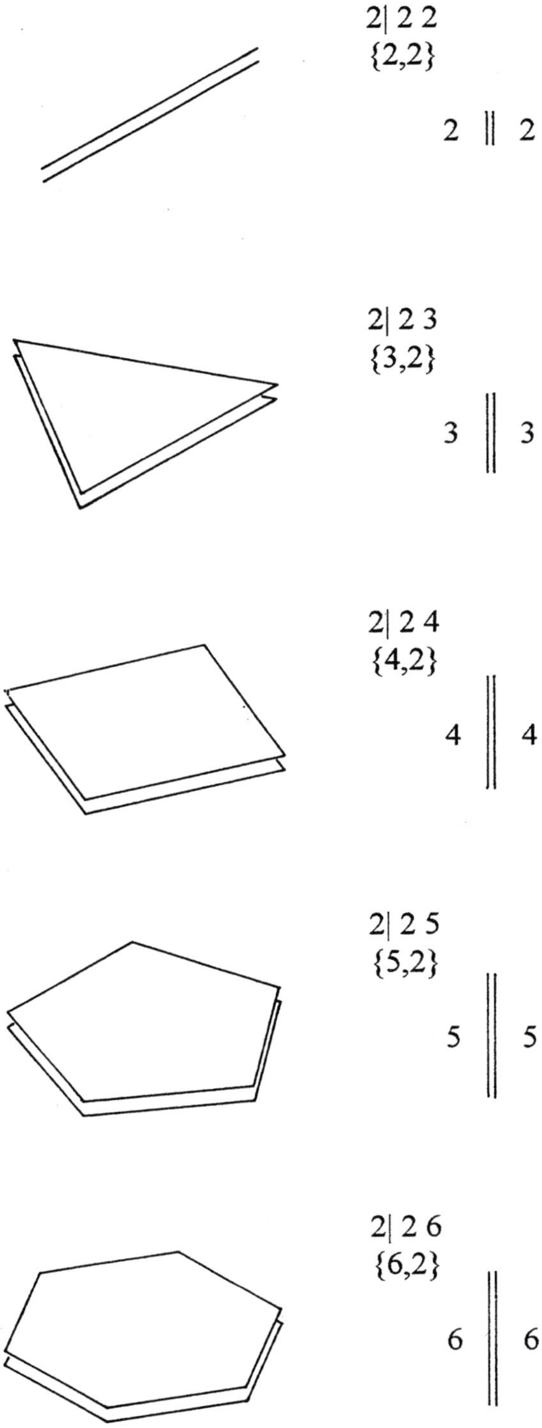

Fig.51: Regular Dihedra

MULTIPLE DIGONS

If the dihedra are to be accepted as regular polyhedra in their own right, they need to have duals, i.e. further regular polyhedra at the other end of a truncation series.

Theoretically these are {2,n}, consisting of multiple bundles of digon faces, meeting at two opposing vertices: n| 2 2.

These are shown in Figure 52 and described in Table 21 below, but can be more clearly appreciated in terms of a dihedral kaleidoscope of (2 2 n) Möbius triangles.

This takes the form of a sphere divided up by an equator with 2n lines of longitude spaced equally around it. When n is even these pair up.

Symbol		V	E	F	
2| 2 2	{2,2}	2	2	2	‖
Double Digon					
3| 2 2	{2,3}	2	3	3	‖
Triple Digon					
4| 2 2	{2,4}	2	4	4	‖
Quadruple Digon					
5| 2 2	{2,5}	2	5	5	‖
Quintuple Digon					
6| 2 2	{2,6}	2	6	6	‖
Sextuple Digon					

Table 21: Multiple Digons

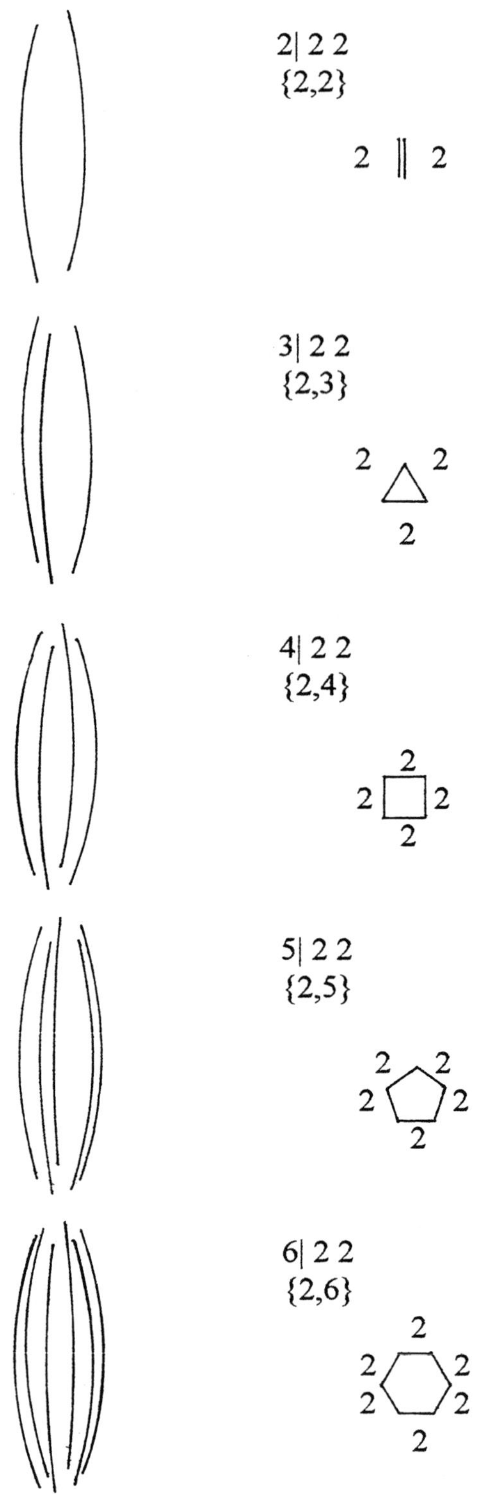

Fig.52: Multiple Digons

TRUNCATED DIGONS: PRISMS

Whilst a bunch of multiple digons may have no volume or even area, making them the most primitive 'polyhedra' we have examined, the results of truncating them are considerably less so.

The theoretical truncation of a digon {2} face is a square {4}, so that what we get as t{2,n} are the prisms consisting of two {n} faces and n squares each: n 2 |2.

These are shown in Figure 53 and described in Table 22 below, and can be seen to have the usual truncated polyhedron isosceles triangle vertex figure.

They also continue to obey Euler's formula.

Symbol		V	E	F
2 2 |2 Double Square	t{2,2}	4	6	2 □ 2 ‖
3 2 |2 Triangle Prism	t{2,3}	6	9	3 □ 2 △
4 2 |2 Square Prism (Cube)	t{2,4}	8	12	4 □ 2 □
5 2 |2 Pentagon Prism	t{2,5}	10	15	5 □ 2 ⬠
6 2 |2 Hexagon Prism	t{2,6}	12	18	6 □ 2 ⬡

Table 22: Truncated Digons

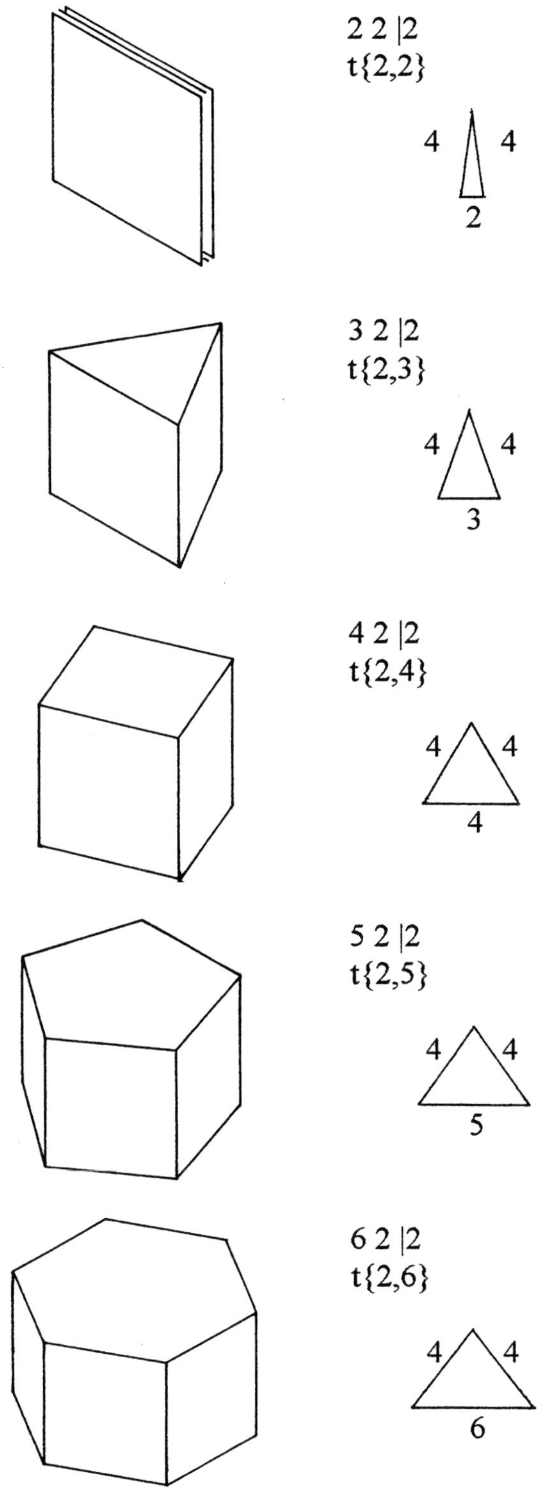

Fig.53: Truncated Digons

TRUNCATED DIHEDRA

Truncating the other member of our dual pairs, the dihedra we started with, we get further dihedra t{n,2}, consisting of two truncated polygons t{n} back to back, with every other edge made double by the inclusion of a digon face there: 2 2 |n.

These are shown in Figure 54 and described in Table 23 below. Again they can be seen to have the usual truncated polyhedron isosceles triangle vertex figure, although very slender on account of the digon {2} face included. This should have a dual edge length of zero, but is shown here with a minimal length for clarity.

Symbol		V	E	F
2 2 |2	t{2,2}	4	2x2 2x1	2 □ 2 ‖
Double Square				
2 2 |3	t{3,2}	6	3x2 3x1	2 ⬡ 3 ‖
Double Hexagon				
2 2 |4	t{4,2}	8	4x2 4x1	2 ○ 4 ‖
Double Octagon				
2 2 |5	t{5,2}	10	5x2 5x1	2 ○ 5 ‖
Double Decagon				
2 2 |6	t{6,2}	12	6x2 6x1	2 ○ 6 ‖
Double Dodecagon				

Table 23: Truncated Dihedra

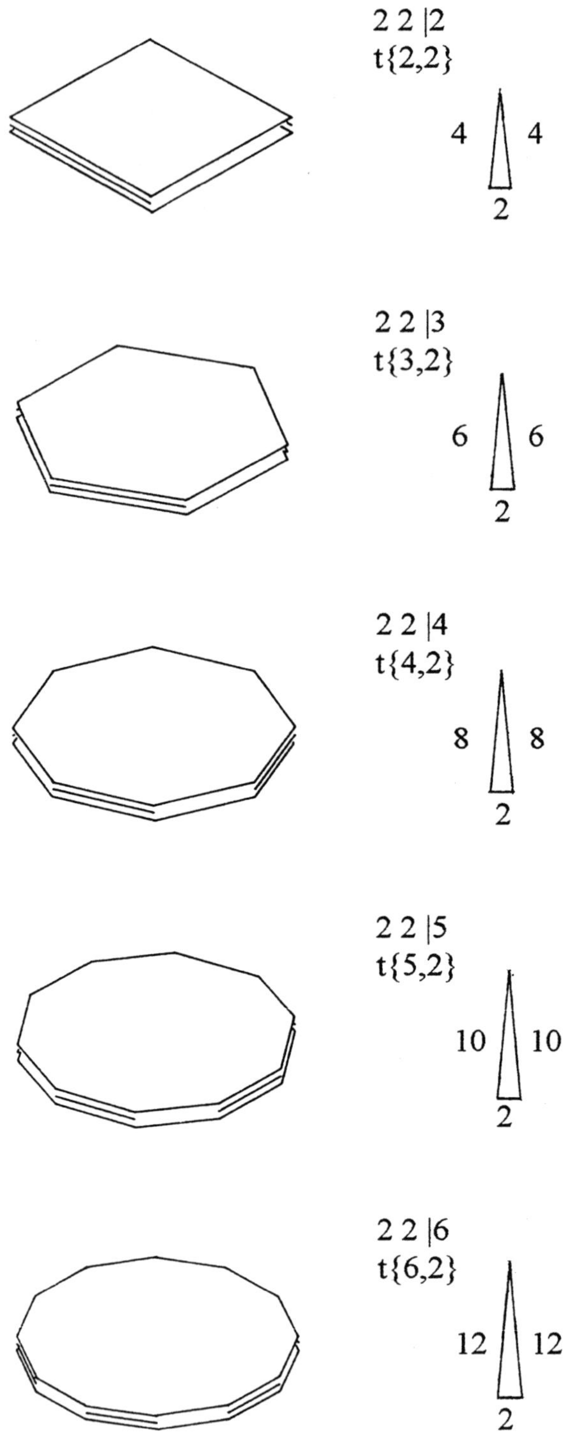

Fig.54: Truncated Dihedra

QUASI-REGULAR DIHEDRA

Continuing the truncation process onwards from either of our truncated forms, we can obtain a quasi-regular dihedron at the centre of each truncation series: 2| n 2, essentially the same as the parent dihedra.

These dihedra again have length and area but no volume, consisting of two parent polygons back to back, but now joined with n double edges all round, containing the n extra digon faces.

Here if we again show the digon {2} faces with a minimal rather than zero dual edge length, we can construct meaningful rectangular vertex figures, as shown in Figure 55. These quasi-regular dihedra are further described in Table 24 below.

Symbol		V	E	F
2| 2 2	{2} {2}	2	2x2	2 ‖ 2 ‖
Double Digon				
2| 3 2	{2} {3}	3	3x2	2 △ 3 ‖
Double Triangle				
2| 4 2	{2} {4}	4	4x2	2 □ 4 ‖
Double Square				
2| 5 2	{2} {5}	5	5x2	2 ⬠ 5 ‖
Double Pentagon				
2| 6 2	{2} {6}	6	6x2	2 ⬡ 6 ‖
Double Hexagon				

Table 24: Quasi-Regular Dihedra

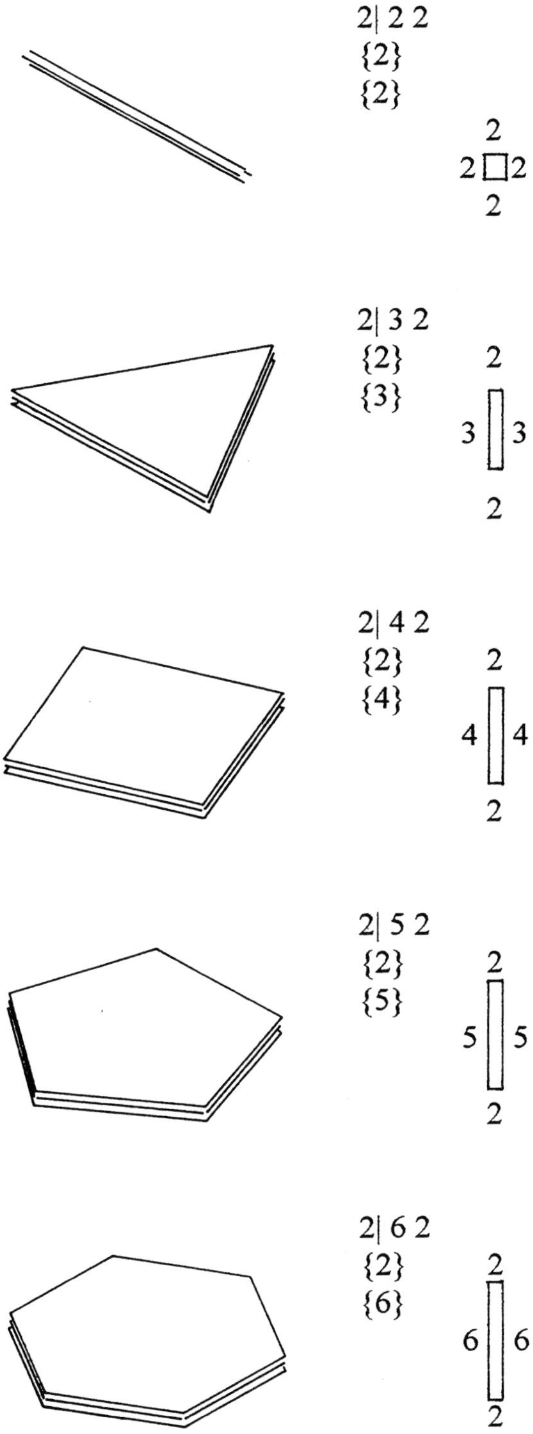

Fig.55: Quasi-Regular Dihedra

EVEN FACED DIHEDRA: PRISMS

We have now seen dihedral symmetry family equivalents of all members of the normal polyhedral symmetry families' first truncation series.

Taking the quasi-regular dihedra we can produce a first truncation in a second series which will be equivalent to the even faced polyhedra: 2 2 n|.

These even faced dihedra are shown in Figure 56 and described in Table 25 below.

They consist of prisms with two truncated polygon faces {2n} plus 2n square faces that are alternately truncated digons and new truncating faces. They have the usual even faced polyhedron triangular vertex figure.

Symbol		V	E	F
2 2 2| Square Prism (Cube)	$t\begin{Bmatrix}2\\2\end{Bmatrix}$	8	12	2 □ 2 □ 2 □
2 2 3| Hexagon Prism	$t\begin{Bmatrix}2\\3\end{Bmatrix}$	12	18	2 ⬡ 3 □ 3 □
2 2 4| Octagon Prism	$t\begin{Bmatrix}2\\4\end{Bmatrix}$	16	24	2 ○ 4 □ 4 □
2 2 5| Decagon Prism	$t\begin{Bmatrix}2\\5\end{Bmatrix}$	20	30	2 ○ 5 □ 5 □
2 2 6| Dodecagon Prism	$t\begin{Bmatrix}2\\6\end{Bmatrix}$	24	36	2 ○ 6 □ 6 □

Table 25: Even Faced Dihedra

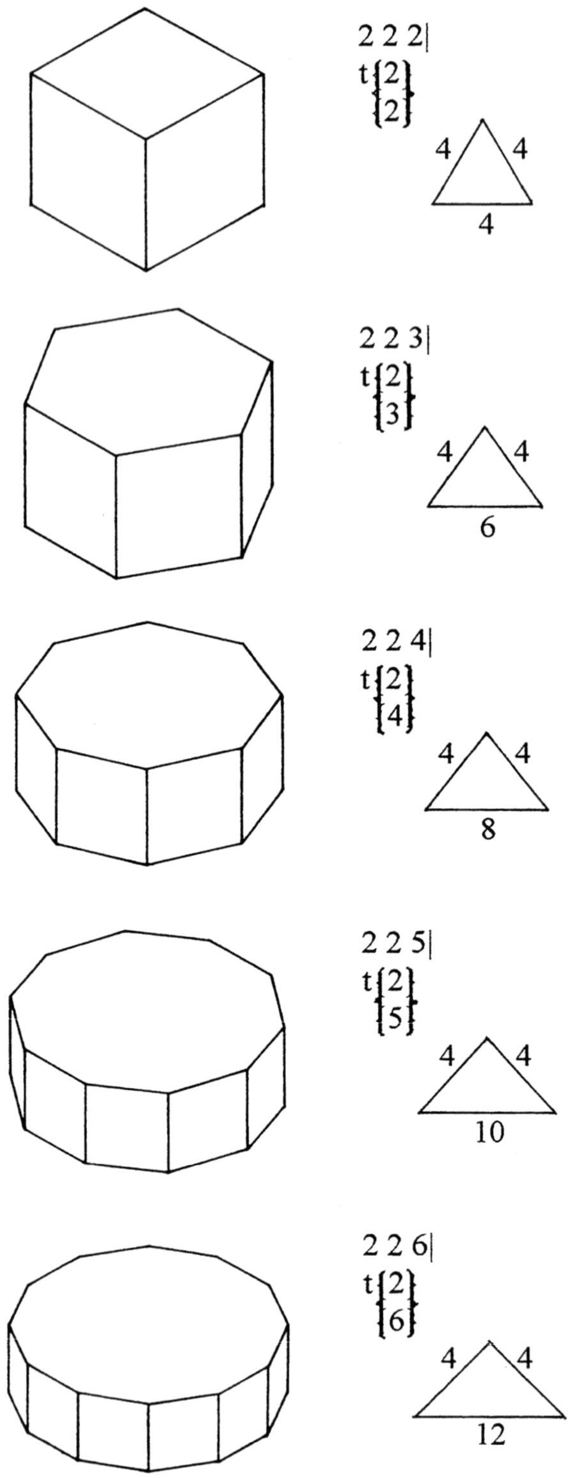

Fig.56: Even Faced Dihedra

RHOMBIDIHEDRA: PRISMS

Continuing this second series truncation process leads ultimately to a dual consisting of n rhomb faces, n of which meet at two vertices and two of which meet at n vertices; essentially another multiple set of bunched digons (where each digon is an elongated rhomb), but this time with extra vertices at the equatorial positions.

En route to this we can find a further member of the second series, the equivalent of the rhombihedra, consisting of two {n} faces, n square faces and n double edged digon faces: 2 n |2. These have the usual rhombihedron isosceles trapezoid vertex figure consisting of two squares joining together {n} and a digon {2} and are shown in Figure 57 and described in Table 26.

Symbol		V	E	F
2 2 |2 Double Square	$r\begin{Bmatrix}2\\2\end{Bmatrix}$	4	2x2 4x1	2 ‖ 2 ‖ 2 □
2 3 |2 Triangle Prism	$r\begin{Bmatrix}2\\3\end{Bmatrix}$	6	3x2 6x1	2 △ 3 ‖ 3 □
2 4 |2 Square Prism (Cube)	$r\begin{Bmatrix}2\\4\end{Bmatrix}$	8	4x2 8x1	2 □ 4 ‖ 4 □
2 5 |2 Pentagon Prism	$r\begin{Bmatrix}2\\5\end{Bmatrix}$	10	5x2 10x1	2 ⬠ 5 ‖ 5 □
2 6 |2 Hexagon Prism	$r\begin{Bmatrix}2\\6\end{Bmatrix}$	12	6x2 12x1	2 ⬡ 6 ‖ 6 □

Table 26: Rhombidihedra

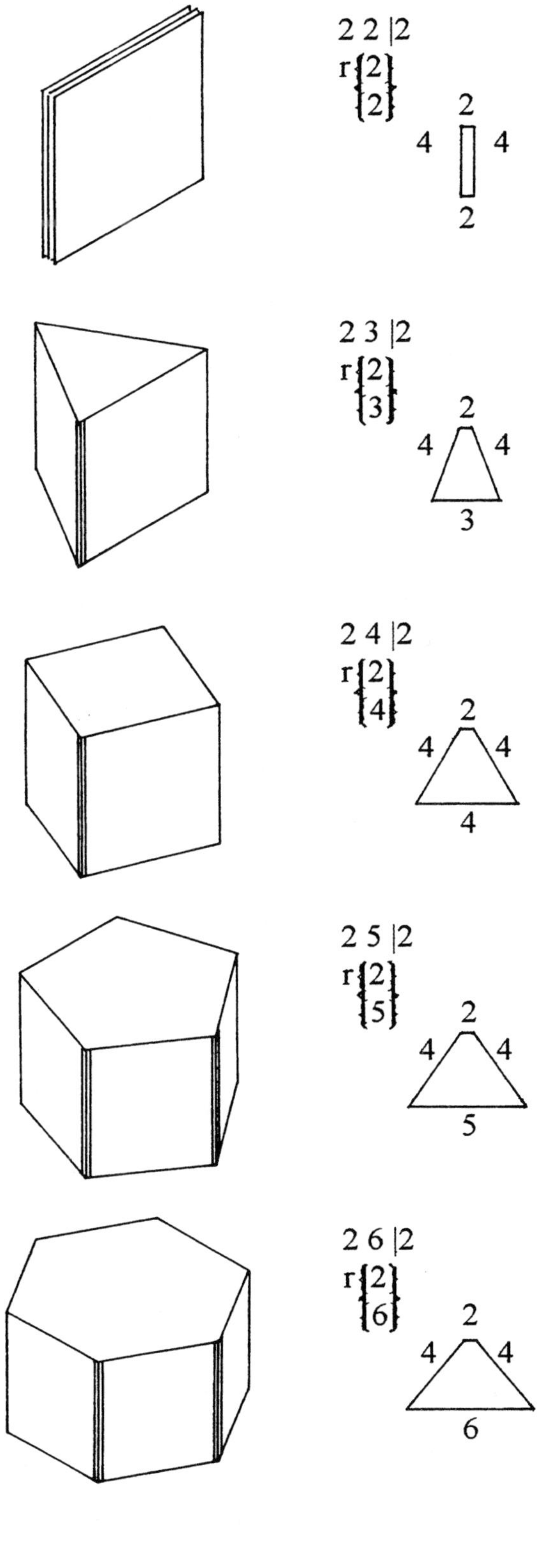

Fig.57: Rhombidihedra

SNUB DIHEDRA: ANTIPRISMS

Taking the rhombidihedra we have just examined, we can remove the square faces, apply the characteristic snub twist and replace them each with two triangles to get snub dihedra completing our second series.

These have two {n} faces, 2n triangle faces and n double edged digon faces and are known as antiprisms: |2 2 n. They have the usual irregular pentagon vertex figure of snubs and are shown in Figure 58 and described in Table 27.

Interestingly the first two of this infinite set are actually a tetrahedron and an octahedron respectively, whilst the fourth member is the central section of an icosahedron. We have come full circle.

<table>
<tr><th>Symbol</th><th></th><th>V</th><th>E</th><th>F</th></tr>
<tr><td>|2 2 2</td><td>$s\begin{Bmatrix}2\\2\end{Bmatrix}$</td><td>4</td><td>2x2
6x1</td><td>2 ‖
2 ‖
4 △</td></tr>
<tr><td colspan="5">Digon Antiprism (Tetrahedron)</td></tr>
<tr><td>|2 2 3</td><td>$s\begin{Bmatrix}2\\3\end{Bmatrix}$</td><td>6</td><td>3x2
9x1</td><td>2 △
3 ‖
6 △</td></tr>
<tr><td colspan="5">Triangle Antiprism (Octahedron)</td></tr>
<tr><td>|2 2 4</td><td>$s\begin{Bmatrix}2\\4\end{Bmatrix}$</td><td>8</td><td>4x2
12x1</td><td>2 □
4 ‖
8 △</td></tr>
<tr><td colspan="5">Square Antiprism</td></tr>
<tr><td>|2 2 5</td><td>$s\begin{Bmatrix}2\\5\end{Bmatrix}$</td><td>10</td><td>5x2
15x1</td><td>2 ⬠
5 ‖
10 △</td></tr>
<tr><td colspan="5">Pentagon Antiprism</td></tr>
<tr><td>|2 2 6</td><td>$s\begin{Bmatrix}2\\6\end{Bmatrix}$</td><td>12</td><td>6x2
18x1</td><td>2 ⬡
6 ‖
12 △</td></tr>
<tr><td colspan="5">Hexagon Antiprism</td></tr>
</table>

Table 27: Snub Dihedra

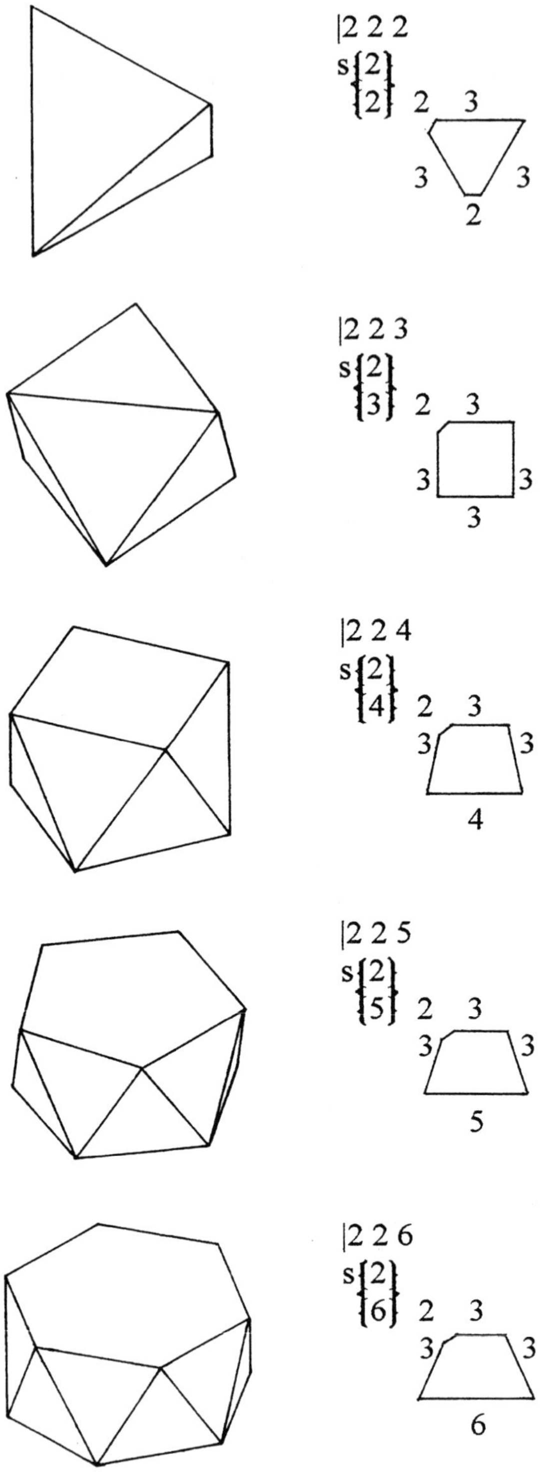

Fig.58: Snub Dihedra

QUASITRUNCATED DIHEDRA

It seems the dihedral symmetry families can produce an equivalent for every one of the normal polyhedra, so naturally we tend to ask what about quasitruncations?

Because a digon is a self-inverse polygon (1/2 + 1/2 = 1), its quasitruncation is the same as its truncation, a single square. We therefore only need to consider the quasitruncation of the {n} constituents of these symmetry families.

Firstly we get the quasitruncated dihedra, t{n',2}, consisting of two t{n'} faces back to back, joined together by n double edges including digons, as shown in Figure 59 and described in Table 28 below: 2 2 |n'.

Symbol		V	E	F
2 2 |2/1	t{2',2}	4	2x2 2x1	2 □ 2 ‖
Double Square				
2 2 |3/2	t{3',2}	3x2	3x2 3x1	2 △ 3 ‖
Double Triangle				
2 2 |4/3	t{4',2}	8	4x2 4x1	2 ✴ 4 ‖
Double Octagram				
2 2 |5/4	t{5',2}	5x2	5x2 5x1	2 ★ 5 ‖
Double Pentagram				
2 2 |6/5	t{6',2}	12	6x2 6x1	2 ✹ 6 ‖
Double Dodecagram				

Table 28: Quasitruncated Dihedra

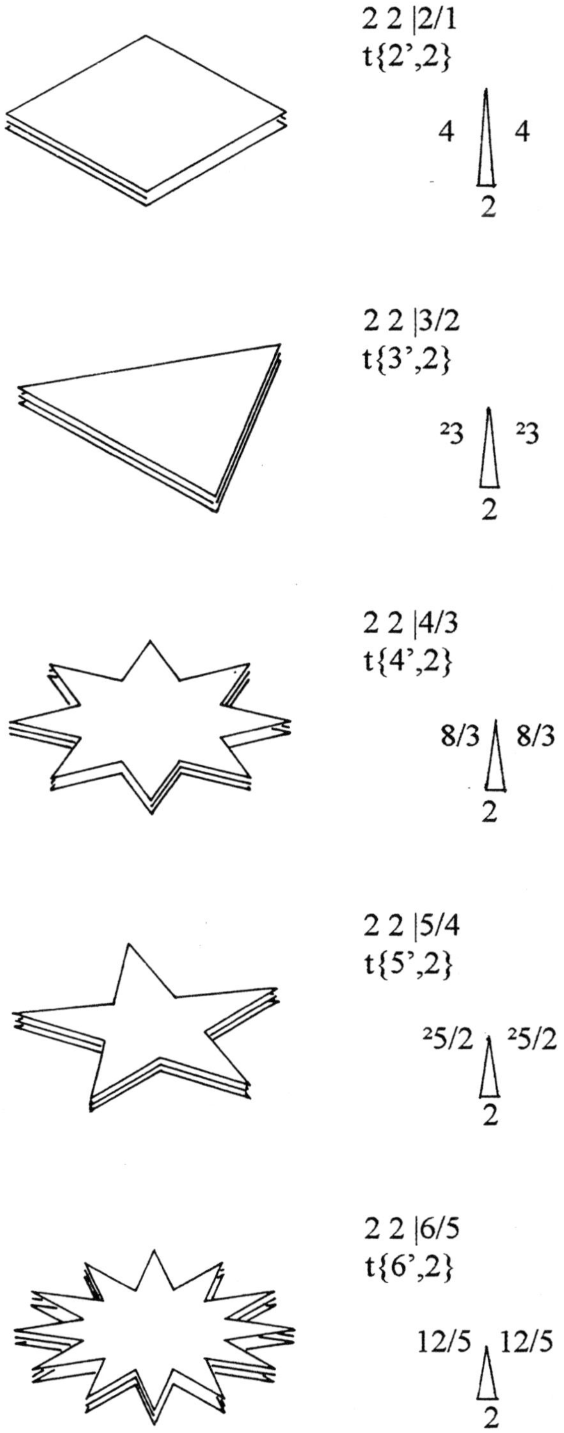

Fig.59: Quasitruncated Dihedra

QUASI EVEN FACED DIHEDRA

The second quasitruncation we can consider is that of the quasi-regular dihedra to produce quasi even faced dihedra. These are prismatic forms of the quasitruncated dihedra just seen on the previous page.

They have similar vertex figures to the even faced dihedra, but with two squares and one quasitruncated {n} at each vertex: 2 2 n'|. These quasi even faced dihedra are shown in Figure 60 and described in Table 29 below.

Some of these are star prisms, members of families of star dihedra, which we shall consider in their own right as an appendix at the end of the next volume dealing with the Star and Cross Polyhedra.

Symbol	V	E	F
2 2 2/1| $t\begin{Bmatrix}2\\2'\end{Bmatrix}$ Square Prism (Cube)	8	12	2 □ 2 □ 2 □
2 2 3/2| $t\begin{Bmatrix}2\\3'\end{Bmatrix}$ Double Triangle Prism	6x2	9x2	2 △ 3 □ 3 □
2 2 4/3| $t\begin{Bmatrix}2\\4'\end{Bmatrix}$ Octagram Prism	16	24	2 ✴ 4 □ 4 □
2 2 5/4| $t\begin{Bmatrix}2\\5'\end{Bmatrix}$ Double Pentagram Prism	10x2	15x2	2 ★ 5 □ 5 □
2 2 6/5| $t\begin{Bmatrix}2\\6'\end{Bmatrix}$ Dodecagram Prism	24	36	2 ✹ 6 □ 6 □

Table 29: Quasi Even Faced Dihedra

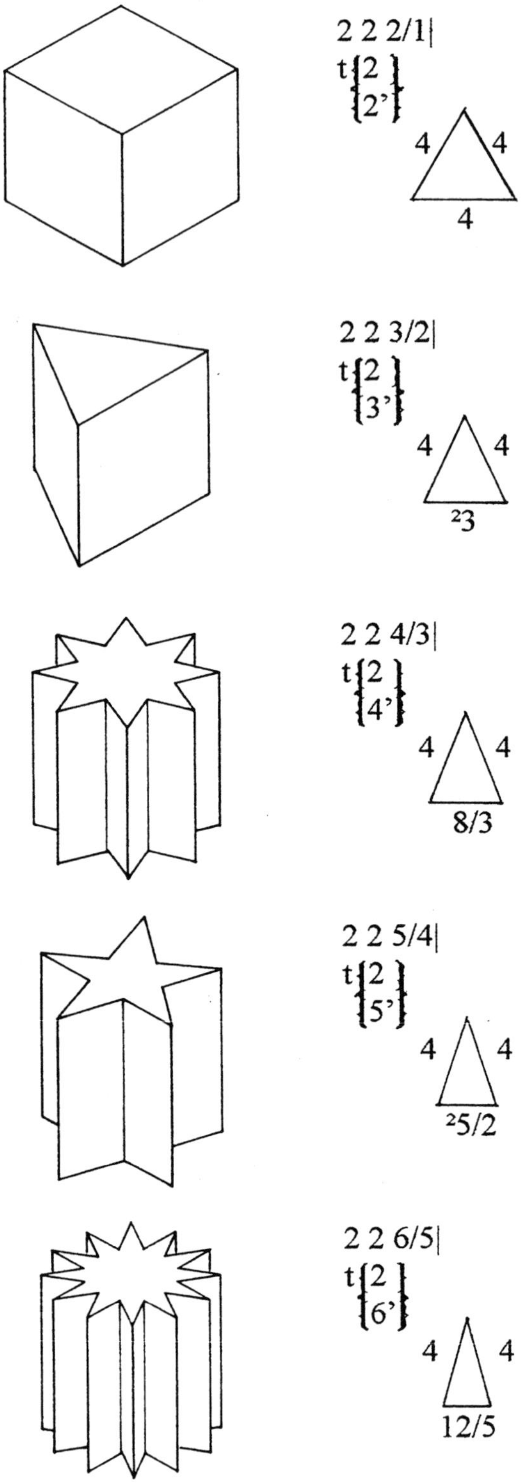

Fig.60: Quasi Even Faced Dihedra

SUMMARY

Except for the addition of the dihedra, the polyhedra we have examined in this volume have run exactly in parallel with the tilings we observed in the first half of the previous volume '*Incomplete Tilings*'.

We first looked at three basic symmetry families of polyhedra generated by three different Möbius triangles (3 2 3), (3 2 4) and (3 2 5). These included simple truncated forms, snubs and the slightly more complex quasitruncated forms for each family.

Schwarz triangles, consisting of pairs of these Möbius triangles back to back, were then examined leading to further polyhedral forms in five sub-families. Some of these were simply double polyhedra, others much more interesting, whilst many gave us hints of further symmetry families for the next volume to investigate.

The generating triangles of these three symmetry families and their five related sub-families, essentially the simpler polyhedra, are shown in Figure 61, along with a typical generating triangle for the even simpler dihedral families (2 n 2).

Coxeter et al. in their famous 1953 paper '*Uniform Polyhedra*' include members from all these families except one, the (3 4/2 3). Because {4/2} is both compound and has zero area, it was not considered a legitimate polygon for inclusion. However the inclusion in the present volume of such compound faces, along with the double polygons resulting from the truncation of even denominator polygons, has hopefully now been justified by the results obtained. This has effectively filled the holes in the standard classification that previously remained curiously empty.

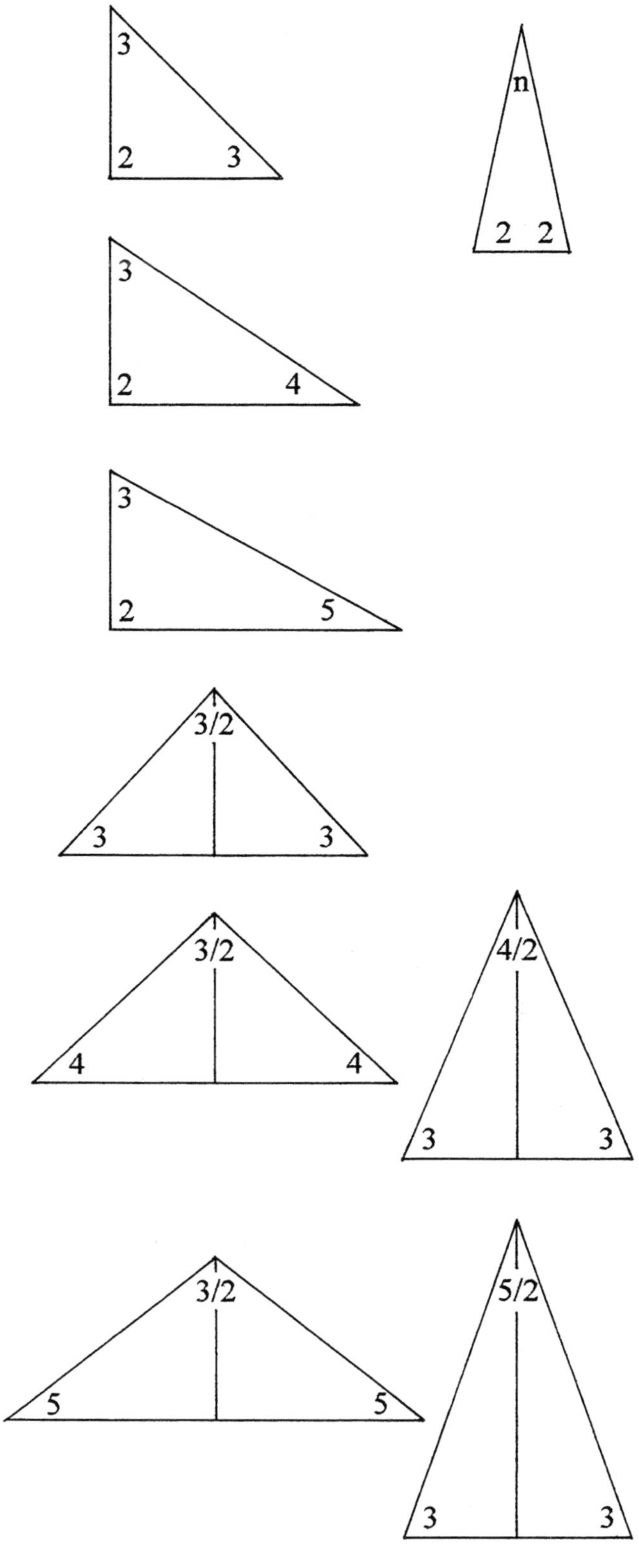

Fig.61: Generating Triangles

SO WHAT FOLLOWS?

We are now just over halfway through the polyhedra according to Coxeter's standard classification. We have encountered 40 of the polyhedra generally acknowledged as 'uniform', along with a good many more that perhaps deserve the title.

Many of these previously disregarded forms do indeed consist of nothing more than other forms inscribed or superimposed, but we have also seen several that can only be disregarded on account of their containing compound or double polygon faces.

It is the contention of this series that these faces are as valid as any others used to construct uniform assemblies of polygons, and that such assemblies are equally valid as 'uniform' tilings or polyhedra. In the next volume, we will therefore continue with business as usual, using the polygons {4/2}, $\{^{2}3\}$, $\{^{2}4\}$ and $\{^{2}5/2\}$ whenever they arise.

We will also continue to run in parallel with the second half of 'Incomplete Tilings' looking firstly at two superfamilies, (3 4/3 4) and (3 5/3 5), before launching into the heavily hinted at (4 2 4/2), (5 2 5/2) and (3 2 5/2) families of cross and star polyhedra and their ensuing sub-families.

Here we will find the remaining 36 'uniform' polyhedra, again amongst a good many more that also perhaps deserve the title.

Following on from this in subsequent volumes, we will continue on down Figure 62, assembling the polyhedra themselves in aggregate as the closepacks filling three dimensional space, which in their turn, by having fewer or lesser valued polyhedra around an edge, can be folded into four dimensions as the hypersolids!

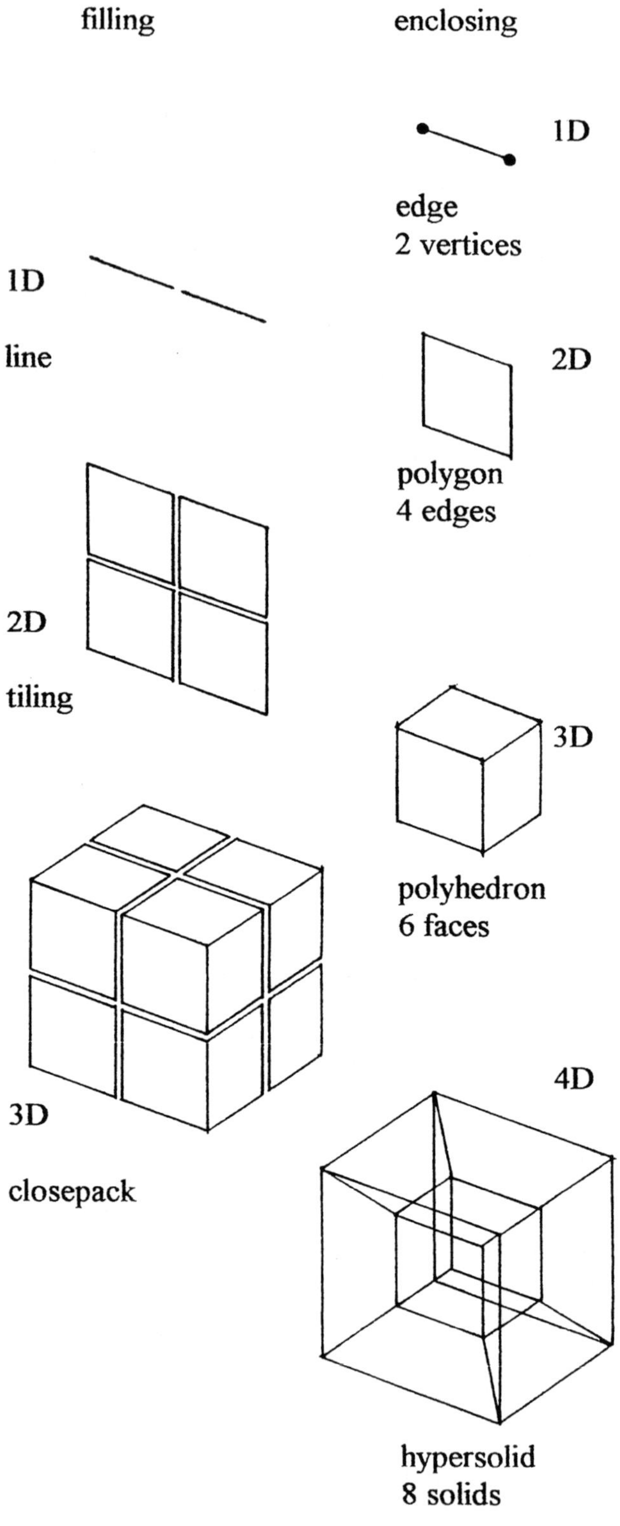

Fig.62: So What Follows?

BIBLIOGRAPHY

Abbott, E A 1962
Flatland
Blackwell

Coxeter, Longuet-Higgins & Miller 1953
Uniform Polyhedra
Phil. Trans. R. Soc. Lond. A246, p401

Coxeter, H S M 1963
Regular Polytopes
Macmillan

Critchlow, K 1969
Order in Space
Thames & Hudson

Cromwell, P R 1997
Polyhedra
Cambridge UP

Cundy, H M & Rollett, A P 1961
Mathematical Models
Oxford

Hargittai, I & M 1994
Symmetry
Shelter

Holden, A 1971
Shapes, Space and Symmetry
Columbia UP

Lakatos, I 1976
Proofs and Refutations
Cambridge UP

Peck, C E 1995
A Taxonomy of Fundamental Polyhedra
Winkworks

Skilling, J 1974
The Complete Set of Uniform Polyhedra
Phil. Trans. R. Soc. Lond. A278, p111

Taylor, P 1990
The Geometry Trip
Nattygrafix

Taylor, P 1995
Additions to the Uniform Polyhedra
Nattygrafix

Taylor, P 1997
The Complete? Polygon
Nattygrafix

Taylor, P 1998
Incomplete Tilings
Nattygrafix

Wenninger, M J 1971
Polyhedron Models
Cambridge UP

Weyl, H 1952
Symmetry
Princeton UP

Williams, R 1972
Natural Structure
Constable

ALSO PUBLISHED BY NATTYGRAFIX:

THE COMPLETE? POLYGON
by Patrick Taylor 1997

being the first part of several comprising
the complete? polyhedra

presents the polygons as building blocks for the tilings, polyhedra etc.; including their generation from a simple vector model along with their truncations, quasitruncations and inverses; regular, star, compound, cross and double polygons; stellations and inscriptions.

60pp A4 paperback including
28 line drawings and 21 tables

ISBN 0 9516701 2 3 £6.00

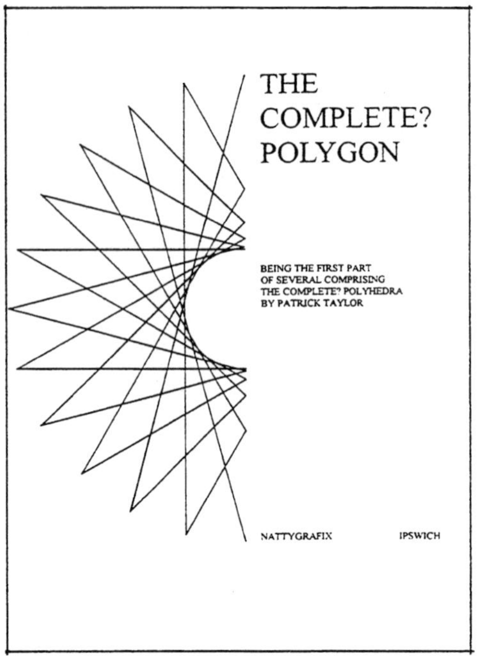

INCOMPLETE TILINGS
by Patrick Taylor 1998

being the second part of several comprising
the complete? polyhedra

presents the tilings as assemblies of polygons that fill the plane; includes truncations, snub tilings, quasitruncations; investigating in total four uniform and seven compound symmetry families using regular, star, cross, compound and double polygons

76pp A4 paperback including
80 line drawings

ISBN 0 9516701 3 1 £6.00